KB263222

에디션 콘텐츠 일러스트

하 랑
도서출판

발행일 : 2025년 08월 30일
출판사 : 하랑출판
주   소 : 서울시 중구 퇴계로28길 8
전   화 : 02-2263-3337

에디션 콘텐츠 일러스트

발행일 : 2025년 08월 30일
출판사 : 하랑출판
주   소 : 서울시 중구 퇴계로28길 8
전   화 : 02-2263-3337

Chapter

computer
arts
163
"It's a graphic designer's job to make the world look like a better place"
__Trevor Jackson, p30
Inside __Jeffrey Bowman • Ubik • Nando Costa
Blend real-life and digital • Jawa & Midwich
Deanne Cheuk • Bertrand Le Pautremat
Hvass&Hannibal • Control type in Flash
Erica Burns • Illustrate song lyrics
From Flickr to Flash • Joshua Davis
Preflighting explained • Mutt Ink
Light trail effects in AE
Inspiration Technique Great Design
The Design Essentials Issue
The new ideas, techniques and inspiration that will make you a better designer
Non-Format: Exclusive interview and cover
28 pages of new Photoshop, Illustrator, InDesign and Flash skills
Iconic advice from Sagmeister, Carson, Brody and Barnbrook
21 design books you must own
The ten best designers you don't know about (yet)
90 minutes of pro Illustrator training on your free disc

computer
arts
163
"It's a graphic designer's job to make the world look like a better place"
__Trevor Jackson, p30
Inside __Jeffrey Bowman • Ubik • Nando Costa
Blend real-life and digital • Jawa & Midwich
Deanne Cheuk • Bertrand Le Pautremat
Hvass&Hannibal • Control type in Flash
Erica Burns • Illustrate song lyrics
From Flickr to Flash • Joshua Davis
Preflighting explained • Mutt Ink
Light trail effects in AE
Inspiration Technique Great Design
The Design Essentials Issue
The new ideas, techniques and inspiration that will make you a better designer
Non-Format: Exclusive interview and cover
28 pages of new Photoshop, Illustrator, InDesign and Flash skills
Iconic advice from Sagmeister, Carson, Brody and Barnbrook
21 design books you must own
The ten best designers you don't know about (yet)
90 minutes of pro Illustrator training on your free disc

BANGERS

BANGERS

CABARET MARTYR
LE CLUB DE LA VIE INIMITABLE
TOUS LES SAMEDIS à 22h
à La Loge Théâtre
2 rue La Bruyère-Paris 9e
M° St-Georges
9€ SEULEMENT
www.cabaretmartyr.com
www.lalogetheatre.fr
lalogetheatre@hotmail.com
Tél. 01 42 82 13 13

CABARET MARTYR
LE CLUB DE LA VIE INIMITABLE
TOUS LES SAMEDIS à 22h
à La Loge Théâtre
2 rue La Bruyère-Paris 9e
M° St-Georges
9€ SEULEMENT
www.cabaretmartyr.com
www.lalogetheatre.fr
lalogetheatre@hotmail.com
Tél. 01 42 82 13 13

The
CHOP SHOP
— NEIGHBORHOOD —
Meat Market
NATURAL MEATS
HONEST DON'S
Since 2009

The
CHOP SHOP
NO 024
WHEN CALLED
IT'S YOUR TIME
FOR SERVICE

The
CHOP SHOP
Grand-Opening!
FIRST ANNUAL
MEAT
& GREET
January 20th 2000 & 10
CHOPSHOP.COM

The
CHOP SHOP
IS A NEW NEIGHBORHOOD
MEAT MARKET with
OLD SCHOOL WITS & REALLY
REALLY SHARP CLEAVERS.
WE PROUDLY SERVE
HONEST DON'S
Local, NATURAL & NUTRITIOUS
GRASS-FED Beef, Lamb, Bison,
Poultry & OTHER Wild Edibles.

DID WE MENTION
THE
FREE
BEER
& HOT DOGS?
SEE YOU AT NOON.
CHOPSHOP.COM

Barbed
Dangerous
UNHOLY ROLLERS v.
RESERVOIR DOLLS
ALLIANT ENERGY CENTER
QUAD SQUAD v.
VAUDEVILLE VIXENS
VETERANS MEMORIAL COLISEUM
TICKETS $10 ADVANCE / $12 DAY OF BOUT
AVAILABLE AT CHA CHA / CAPITOL CITY TATTOO
LAKESIDE PRESS / ONLINE AT MADROLLINDOLLS.COM
YOUNG LIVERS / THE HUSSY

ALL GOOD IS HIDDEN AWAY
Then why
believe
the obvious?

Bare feet
on the ground

Life is too short
to set it on pause
Just hit play

The New York Times Magazine
DECEMBER 28, 2008
Tim Russert
Jean Winston
Philip Agee
George Carlin
Bobby Fischer
Heston
Gill Elder
Mildred Loving
Jim McKay
Eula Mae Doré
Stubbs
THE LIVES THEY LIVED

12
A
LESE
ZEICHEN

ZWEIT
GEDANKEN

country of origin
Denmark

description
A promotional brochure for Pacesetters, describing their design philosophy and working methods.

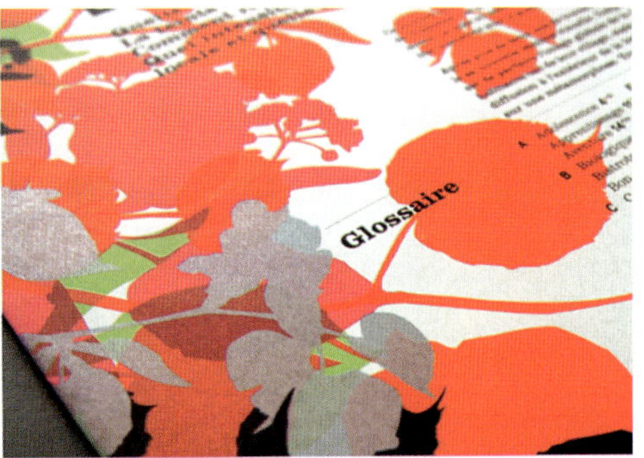

typeface

bf_SubZero

typeface family
bf_SubZero

designer
Guido Schneider

foundry/supplier
brass_fonts cologne

country of origin
Germany

Sinn—
Sisamouth
1935–1975
was a famous
and highly
prolific
Cambodian
singer-
songwriter in
the 1950s
to the 1970s.

1960s
Cambodian
music
scene
Typeface: Sisamouth
105 point / 82 / tomtor studio@
www.tomtor.com
Sinn—
Typeface Sisamouth
200 point / 126 / tomtor studio@
www.tomtor.com
Singer-
Songwriter
in the
1950s

You've really
made the grade

and may God's love be with you
5
4
3

Liftoff
2

description

An invitation for a lecture on
Dutch design, organized by the
Society of Typographic Designers.

serenità + auguri + allegria
prosperità + LEGGERE + CROCCANTI + GUSTOSE +
più
2010

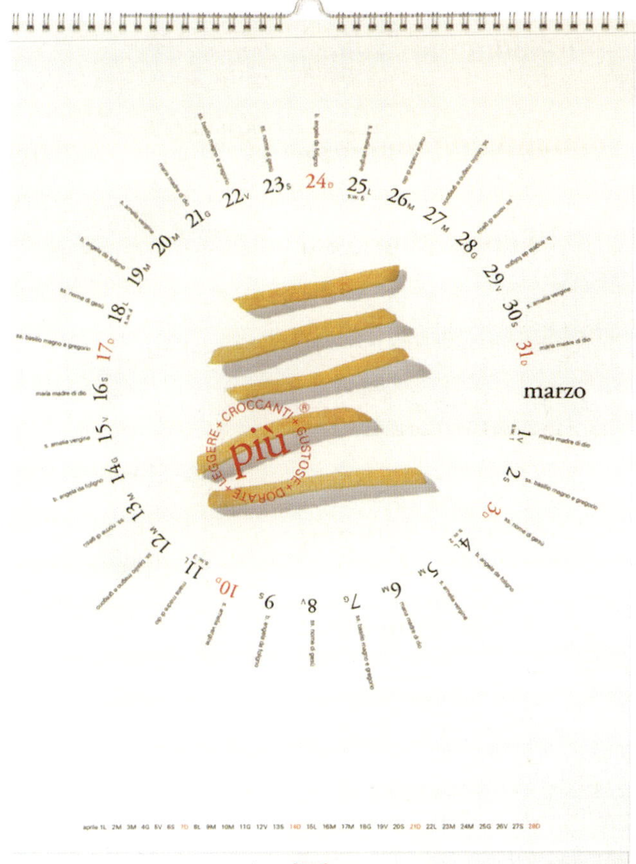
più
CROCCANTI GUSTOSE
LEGGERE DORATE
marzo

più
CROCCANTI GUSTOSE
LEGGERE DORATE
aprile

A wall calendar for German design studio Gaga Design, created in 1997 to commemorate their 10th anniversary. The design incorporates cartographic elements, timezone graphics, and topographic earth maps, picking up on the themes of seasons, climate, leisure, and holidays to celebrate free time rather than work days.

Sample from a "letter lottery," where each student draws lots to design five to six characters for a typeface. One student receives six letters from the instructor; then, based on the formal parameters defined by these six letters, the student creates the characters assigned to them in the letter lottery. Finally, they erase the first six letters and hand the newly designed ones to the next student, who repeats the process until each student has designed their allotted characters and the typeface is complete.

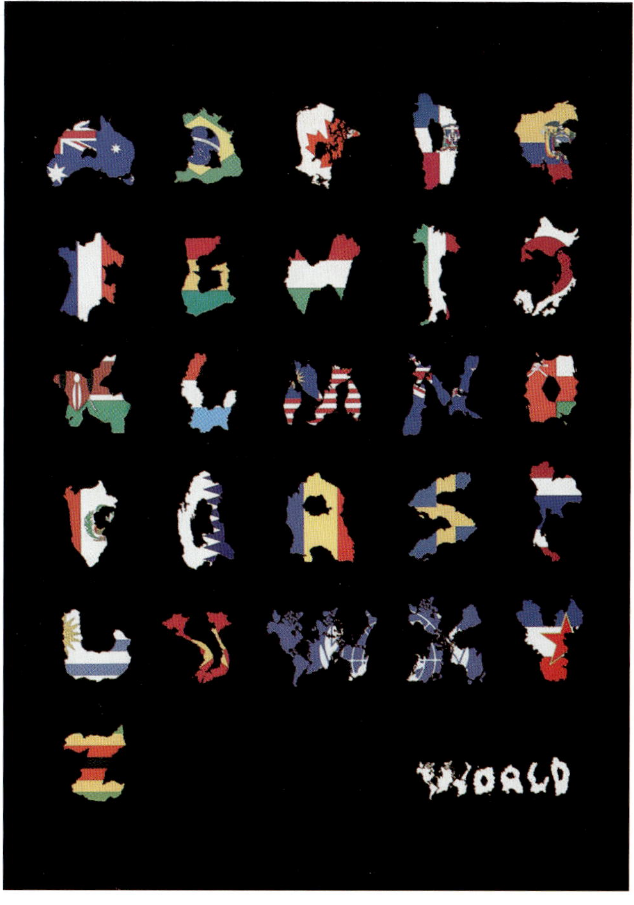

ECOLOGY
IS GREEN
AND IT'S ALSO
BLUE

WE WANT
+ H2O
WE WANT
- CO2

Estrella
Damm
BIENVENIDO AL
MEDITERRANEO

typeface

CIA Humdrum

typeface family
The CIA Compendium

designer
Jens Gehlhaar

foundry/supplier
Jens Gehlhaar

country of origin
USA

Dušan Jelesijević

U godina, 2007
Diplomski rad

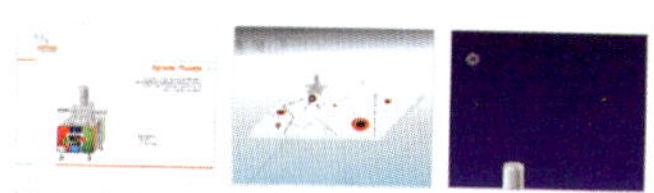

CD sadrži:

Knjigu standarda (PDF)
Seriju plakata (PDF)
Font ARTVOD (TTF)
Prezentaciju fonta ARTVOD (PDF)
Interaktivnu prezentaciju (SWF, EHE)
Instalacione programe (Flash Player, Opera)

THE COMMUTE IS SUCH
A DRAIN ON AN INDIVIDUALS BODY.
SPENDING
ALMOST HALF YOUR DAY JUST IN
EVERY MORNING I HAVE TO EXPERIENCE
THE RIDE TO DOWNTOWN TORONTO
WHERE I ATTEND SCHOOL.
I COULD ALWAYS GO TO SHERIDAN, IT'S ONLY A 10MIN DRIVE
MOST OF THE TIME I SIT THERE
STARING OUT THE WINDOW IN A DAZE,
ALWAYS WONDERING
"WHAT IF I WAS STILL IN MY OWN BED?
WARM IN MY COMFORTER WITH
NOT A CARE IN THE WORLD."
BUT NO, THIS IS REALITY
AND WE HAVE TO
FACE THE FACTS.
IF WE WERE JUST TO
SIT AROUND ALL DAY IN BED
WHAT WOULD OUR
LIVES BECOME OF?

FREEDOM
FAMILY LOVE AND SUPPORT
RESPONSIBILITY
SELF RESPECT
SOCIAL AND
MANAGEMENT AND SUC
ACHIEVEME

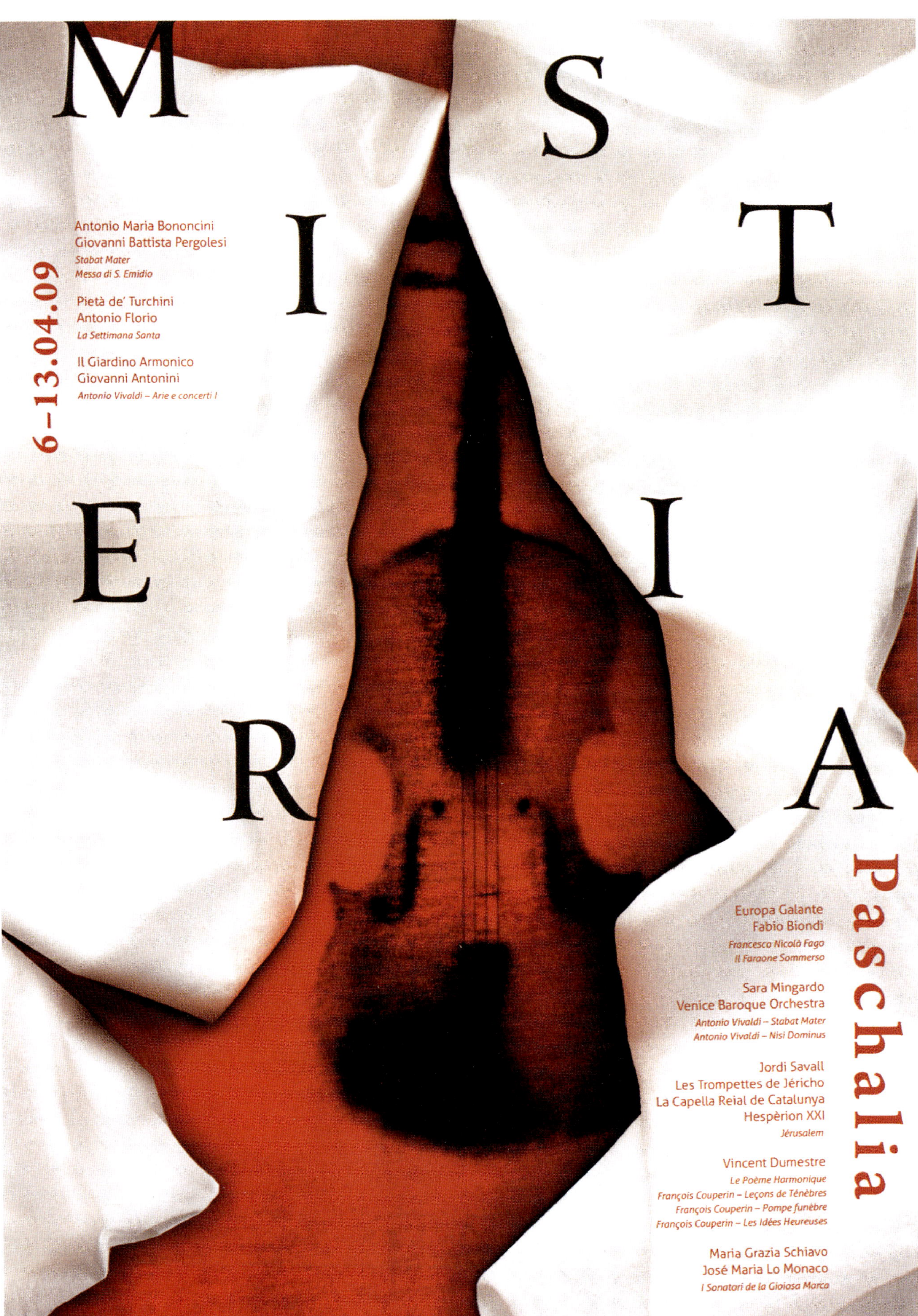
MISTERIA
Paschalia

6–13.04.09

Antonio Maria Bononcini
Giovanni Battista Pergolesi
Stabat Mater
Messa di S. Emidio

Pietà de' Turchini
Antonio Florio
La Settimana Santa

Il Giardino Armonico
Giovanni Antonini
Antonio Vivaldi – Arie e concerti I

Europa Galante
Fabio Biondi
Francesco Nicolò Fago
Il Faraone Sommerso

Sara Mingardo
Venice Baroque Orchestra
Antonio Vivaldi – Stabat Mater
Antonio Vivaldi – Nisi Dominus

Jordi Savall
Les Trompettes de Jéricho
La Capella Reial de Catalunya
Hespèrion XXI
Jérusalem

Vincent Dumestre
Le Poème Harmonique
François Couperin – Leçons de Ténèbres
François Couperin – Pompe funèbre
François Couperin – Les Idées Heureuses

Maria Grazia Schiavo
José Maria Lo Monaco
I Sonatori de la Gioiosa Marca

F6
ProDesign
www.prodesign.ru

ABSOLUT
CREATIFF

ProDesign
www.prodesign.ru

ABSOLUT
VODKA
IMPORTED
ABSOLUT
LIGHT
ProDesign

NIKE

Visual identity and art direction for 'Who's Next', a French fashion show that has become one of the top international events. Fragments of headline text set in Neue Helvetica have been cut up and pasted onto the spreads. These cut-ups give the pages a contemporary punk-aesthetic twist.

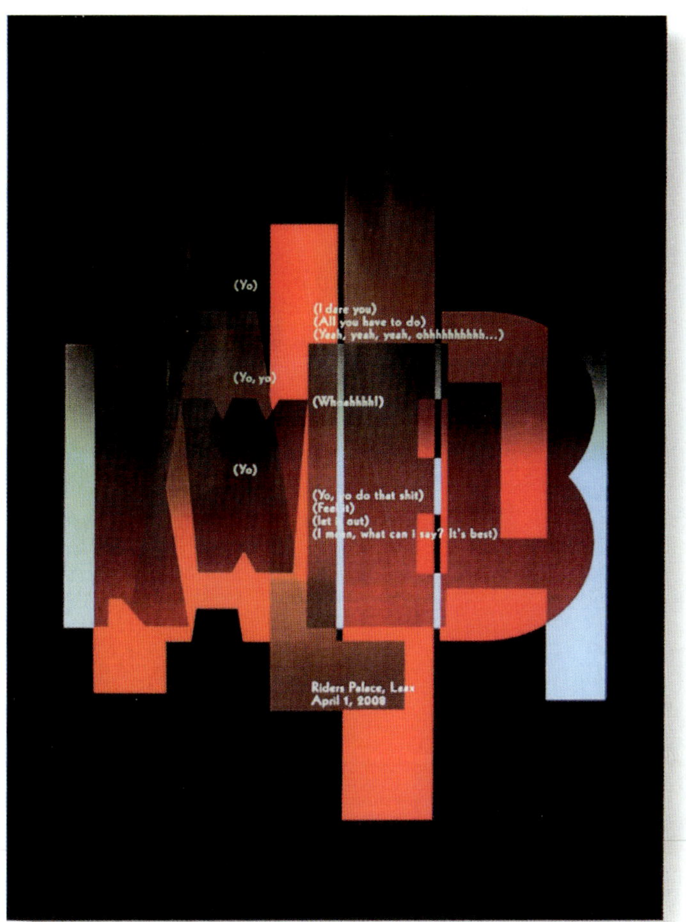

(Yo)
(I dare you)
(All you have to do)
(Yeah, yeah, yeah, ohhhhhhhhhh...)
(Yo, yo)
(Whaahhhh!)
(Yo)
(Yo, yo do that shit)
(Feel it)
(let it out)
(I mean, what can i say? It's best)
Riders Palace, Laax
April 1, 2008

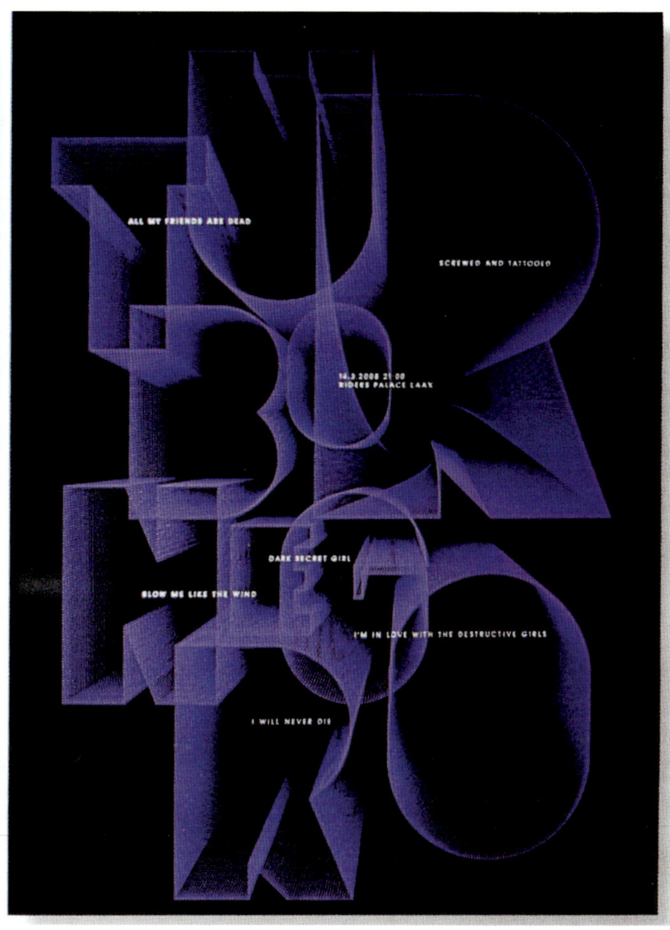

ALL MY FRIENDS ARE DEAD
SCREWED AND TATTOOED
14.3.2008 21.00
RIDERS PALACE LAAX
DARK SECRET GIRL
BLOW ME LIKE THE WIND
I'M IN LOVE WITH THE DESTRUCTIVE GIRLS
I WILL NEVER DIE

Soko.
by
WE MAKE POGO

Soko.
by
NACHO RICCI

Soko.
by
JONATHAN LEDER

BOOST THE XCITEMENT

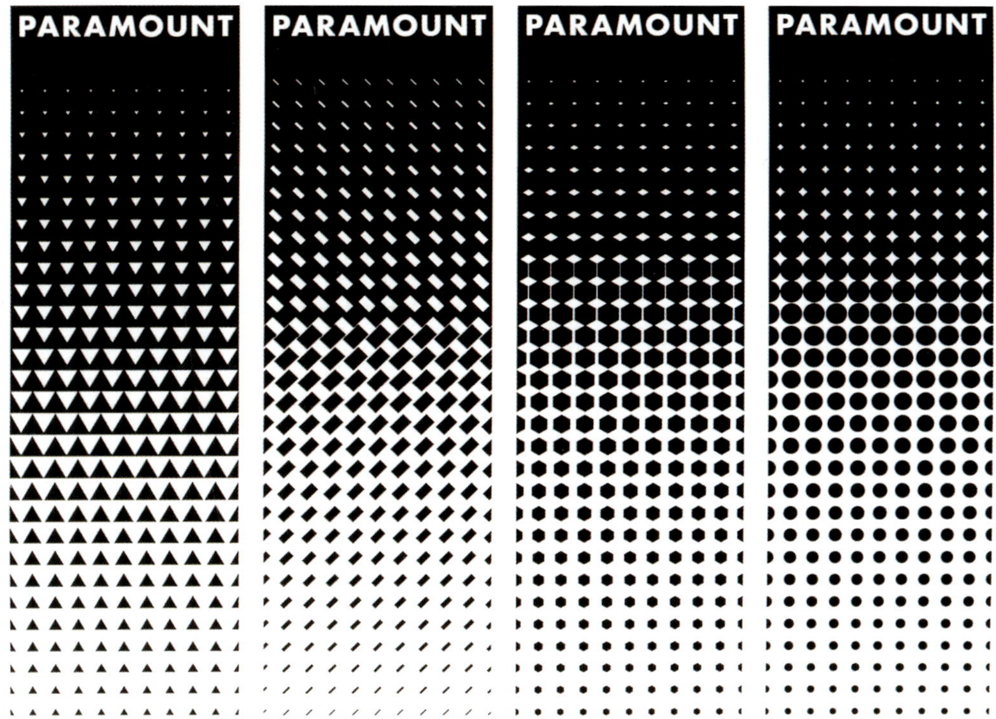

PARAMOUNT
PARAMOUNT
PARAMOUNT
PARAMOUNT

MEETING
PRODUCTIONS
INC.
MEETING
PRODUCTIONS
INC. Quality Event
Planning
MEETING
PRODUCTIONS
INC. Quality Event
Planning

MEETING
PRODUCTIONS
INC. Quality Event
Planning
MEETING
PRODUCTIONS
INC.

MEETING
PRODUCTIONS
INC. Quality Event
Planning

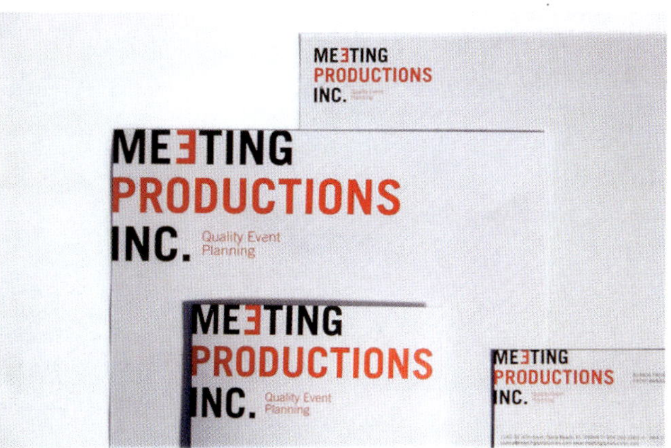

MEETING
PRODUCTIONS
INC.
MEETING
PRODUCTIONS
INC. Quality Event
Planning
MEETING
PRODUCTIONS
INC. Quality Event
Planning
MEETING
PRODUCTIONS
INC.

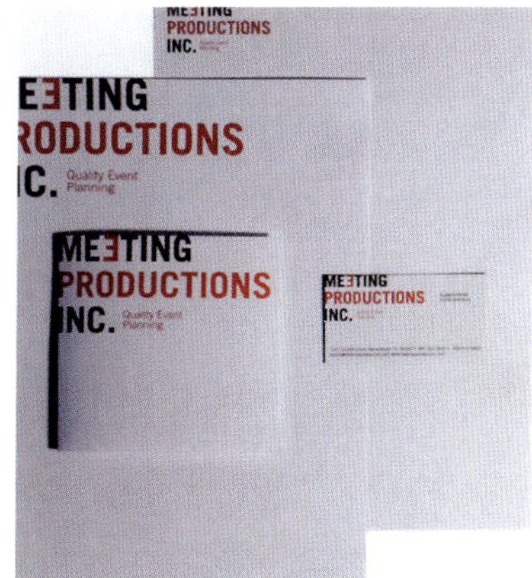

MEETING
PRODUCTIONS
INC.
MEETING
PRODUCTIONS
INC. Quality Event
Planning
MEETING
PRODUCTIONS
INC. Quality Event
Planning
MEETING
PRODUCTIONS
INC.

MEETING
PRODUCTION
INC. Quality Event
Planning

Beginning life as an identity for the typefoundry Dalton Maag, the DaMa logo has been extended into a full typeface. The fat rounded letterforms have a retro quality which is brought up to date by the reduced, minimal shape of each character. The font is only available in one heavy weight, which is generated in both a solid and an outline

Produced for a typefoundry, this brochure highlights a series of bespoke corporate fonts created by it. Large fragments of the fonts are shown against photographic backdrops relevant to each case study. The typography is interwoven with screen grabs of the fonts under construction.

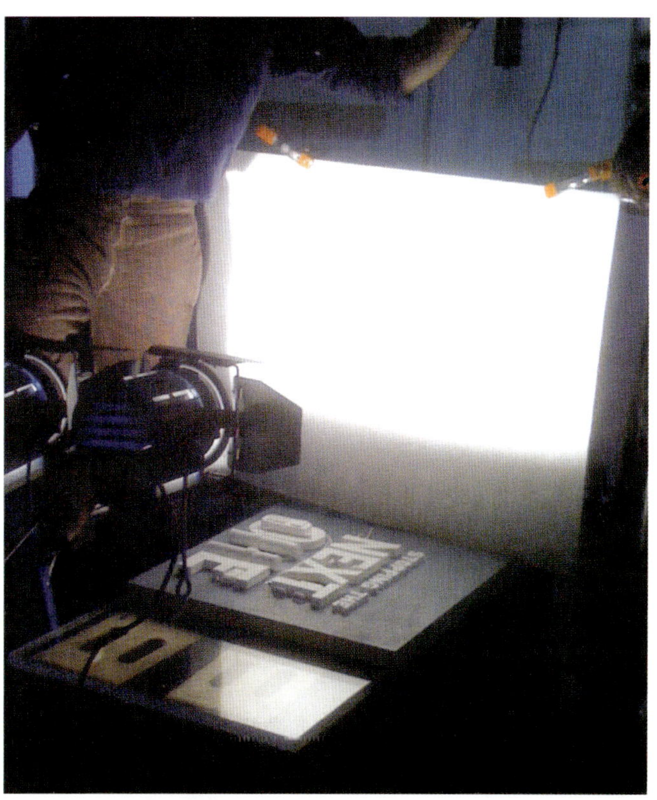

STOPPING THE
NEXT ONE

ARTHUR LUBOW: MAKING ART OUT OF AN ENCOUNTER TOM DUNKEL: SHOOTING UP TO SLOW AGING AMANDA HESSER: PERFECTING THE PANCAKE
The New York Times Magazine
JANUARY 17, 2010
STOPPING THE
NEXT
ONE
OBAMA'S WAR OVER TERROR BY PETER BAKER

A simple folded sheet of high-gloss black paper is enlivened by the application of silver glitter to the custom-drawn typography. The geometric typography on this New Year's card from the London-based design consultancy is constructed from just a circle and a straight line. The design combines a retro Seventies disco aesthetic with a clean modern sensibility.

Based on one of the designer's own typeface designs, these New Year's resolutions have been sprayed onto the studio wall to serve as a semi-permanent reminder for the rest of the year.

Llegeix
Lee
Read
Interactiu
Interactivo
Interactive
Materials
Materiales
Museus
Museos
Museums
Botiga
Tienda
Shop
Producte
Producto
Product
Debats
Debates
Compareix
Comparte
Share
Descobreix
Descubre
Discover
Exposiciones
Expositions
Aprendre
Aprender
Learn
Llegeix
Lee
Read
Tècnic
Técnico
Technique
Llegeix
Lee
Relax
Relaje
Relax
Futur
Futuro
Future
Fes
Haz
Do
Escolta
Escucha
Listen
Disseny
Interactiu
Interactivo
Comunicació visual
Comunicación visual
Visual Communication
Botiga
Tienda
Shop
Relaxa
Relájate
Fes
Haz
Do
Producte
Producto
Product
Disseny
Diseño
Museus
Museos
Debats
Debates
Jocs
Juguetes
Toys
Tallers
Talleres
Workshops
Teoría
Theory
Exposicions
Exposiciones
Exhibitions
Producte
Producto
Prod
Arquitectura
Architecture
Futur
Futuro
Future
Llegeix
Lee
Read
Aprendre
Aprender
Learn
Toca
Touch
Diagnosi
Diagnosis
Hub
Barcelona
DHUB
Interactiu
Interactive
Descobreix
Descubre
Discover
Toca
Touch
Reflex
Hub
Fes
Haz
Do
Museus
Museos
Museums
Diagnosi
Diagnosis
Acció
Acción
Action
Compareix
Com
Share
Escolta
Escucha
Listen
Barcelona
Indumentària i Moda
Indumentaria y Moda
Clothing & Fashion
Relax
Exposicions
Exposiciones
Exhibitions
Descobreix
Descubre
Descobreix
Descubre
Exposicions
Exposiciones
Descobreix
Descubre
Discover
Producte
Producto
Product

NTUA&D
NOTTINGHAM
TRENT
UNIVERSITY
ART
&
DESIGN
BOOK
NOTTINGHAM
TRENT UNIVERSITY

UNCANNY
FIGURES AND
MEAN BODY
BY KIM
SAWCHUK

Q&A

PREFIX PHOTO
Prefix
Photo
35:
Spectral
Light

NTUA&D
STUDEN
STUD
PRES

PREFIX PHOTO
Prefix
Photo
17:
The Last
Photograph

CD
GHI
STU
XYZ

These three record sleeves produced for Ninja Tune/
Smalltown Supersound Records feature a more heavily
computer-rendered CAD approach than is usually
associated with the designer, and combine montaged gritty
photography with slick hard typographic edges.

immaculately designed without losing an ounce of professionalism, we were able to !nd a visual language which helped bring this brand into a new century which was a key element in developing a new signature for Kaare Berntsen. We worked on the development of the new design strategy to implement a fill identity on all levels. From business cards to sales material and exterior and interior surfaces.

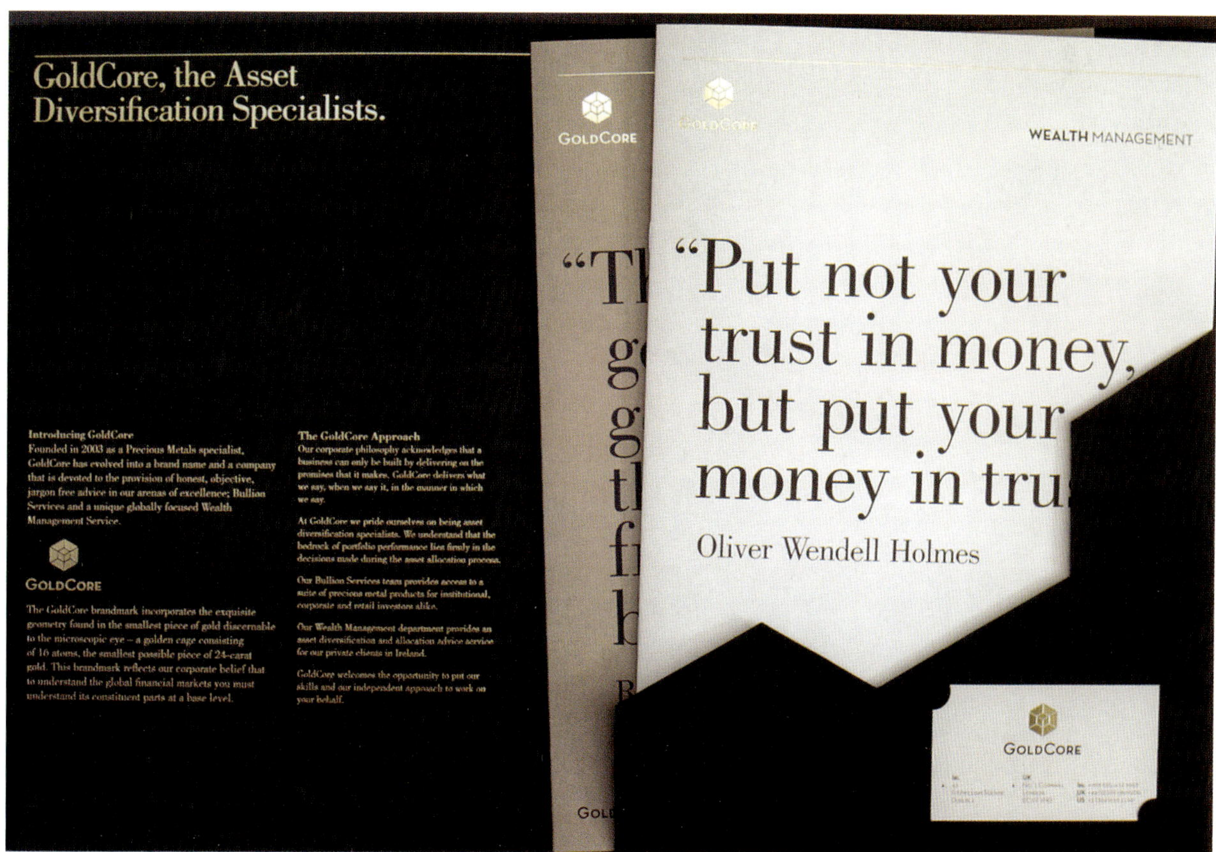
GoldCore, the Asset
Diversification Specialists.

GOLDCORE

WEALTH MANAGEMENT

"Put not your
trust in money,
but put your
money in tru
Oliver Wendell Holmes

Introducing GoldCore
Founded in 2003 as a Precious Metals specialist,
GoldCore has evolved into a brand name and a company
that is devoted to the provision of honest, objective,
jargon free advice in our arenas of excellence; Bullion
Services and a unique globally focused Wealth
Management Service.

GOLDCORE

The GoldCore brandmark incorporates the exquisite
geometry found in the smallest piece of gold discernable
to the microscopic eye – a golden cage consisting
of 16 atoms, the smallest possible piece of 24-carat
gold. This brandmark reflects our corporate belief that
to understand the global financial markets you must
understand its constituent parts at a base level.

The GoldCore Approach
Our corporate philosophy acknowledges that a
business can only be built by delivering on the
promises that it makes. GoldCore delivers what
we say, when we say it, in the manner in which
we say.

At GoldCore we pride ourselves on being asset
diversification specialists. We understand that the
bedrock of portfolio performance lies firmly in the
decisions made during the asset allocation process.

Our Bullion Services team provides access to a
suite of precious metal products for institutional,
corporate and retail investors alike.

Our Wealth Management department provides an
asset diversification and allocation advice service
for our private clients in Ireland.

GoldCore welcomes the opportunity to put our
skills and our independent approach to work on
your behalf.

GOLDCORE

typeface

Splat

typeface family
Splat

designer
Garry Waller

foundry/supplier
Garry Waller

country of origin
UK

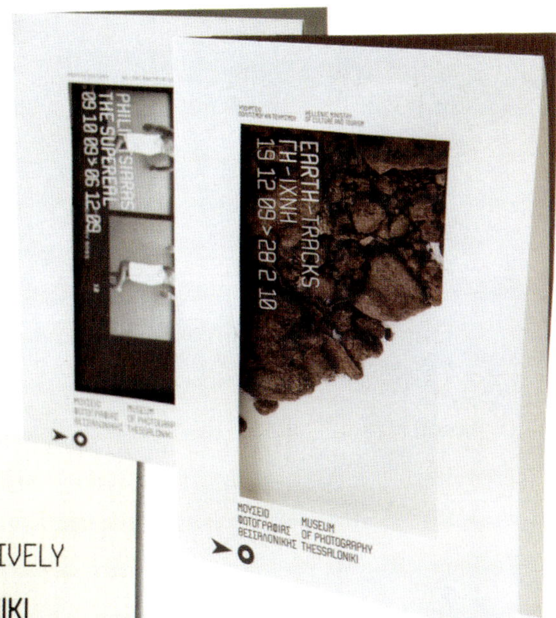

MAKOTO TYPEFACE
IS A DISPLAY FONT
DESIGNED EXCLUSIVELY
FOR THE NEEDS
OF THE THESSALONIKI
PHOTOGRAPHY MUSEUM
0123
4567
890
&$%?€£

.Mene
er,de.
wit.

.M
en
er.
de.
.wit.

.M
enee
r
.de.
.wit.

.M
en
ee
r.de.
.wit
.

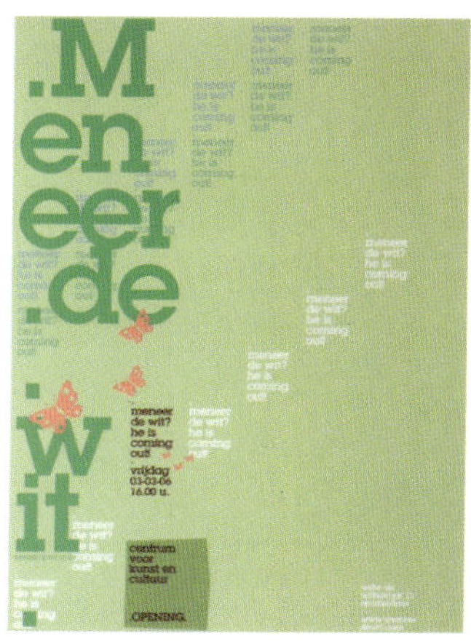

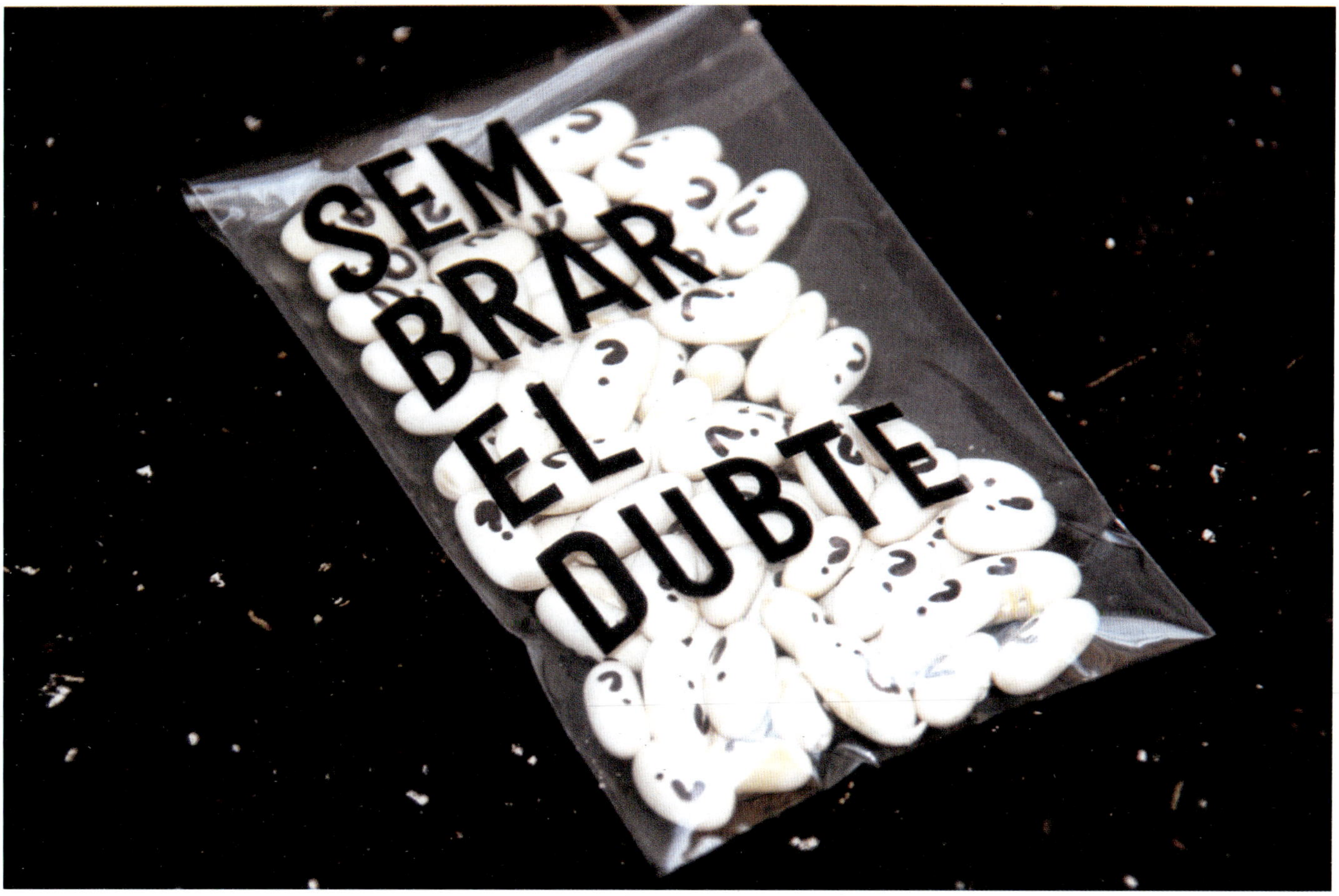

SEM
BRAR
EL
DUBTE

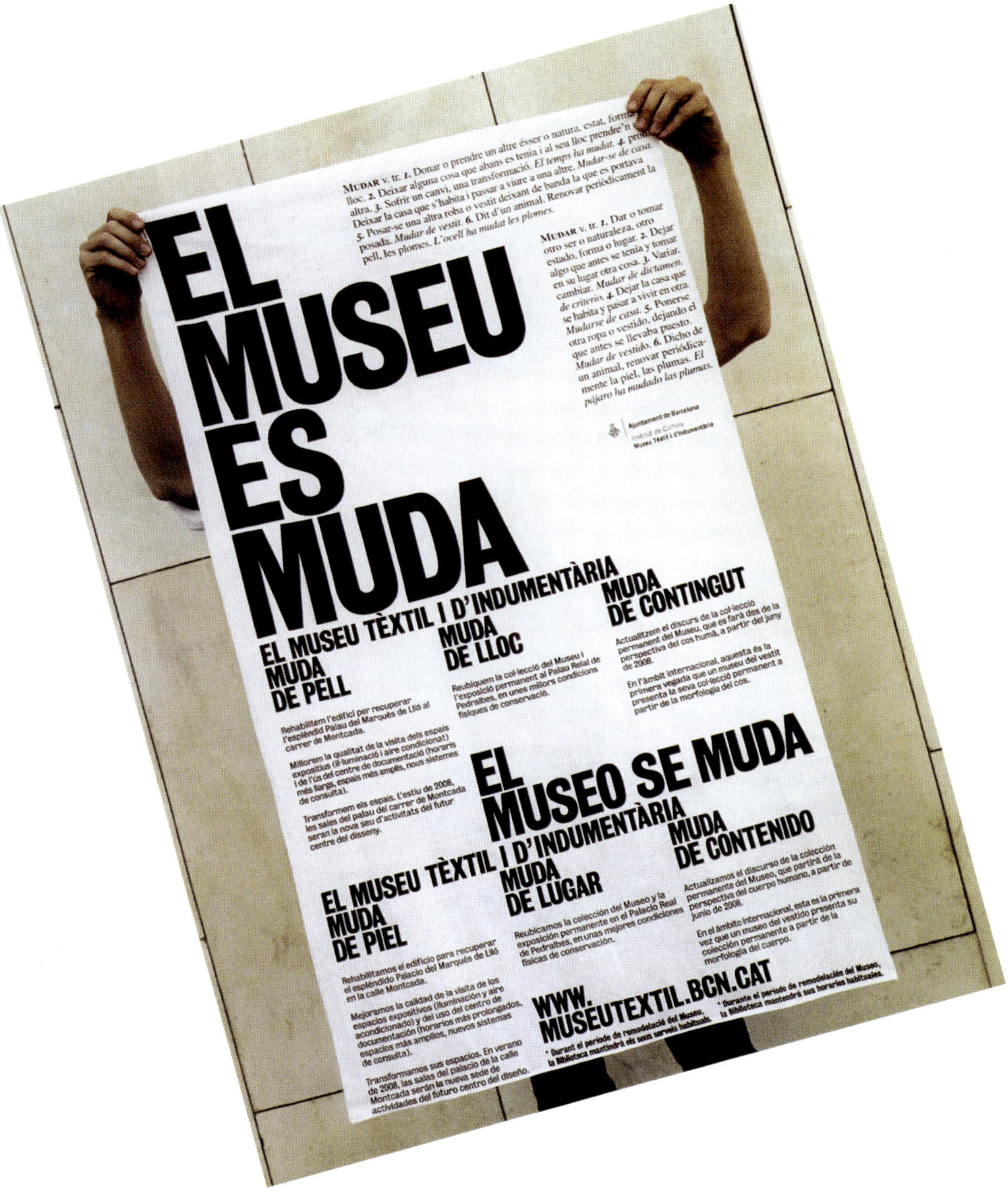
MUDAR v. tr. 1. Donar o prendre un altre ésser o natura, estat, forma, lloc. 2. Deixar alguna cosa que abans es tenia i al seu lloc prendre'n una altra. 3. Sofrir un canvi, una transformació. El temps ha mudat. 4. Deixar la casa que s'habita i passar a viure a una altre. Mudar-se de casa. 5. Posar-se una altra roba o vestit deixant de banda la que es portava posada. Mudar de vestit. 6. Dit d'un animal. Renovar periòdicament la pell, les plomes. L'ocell ha mudat les plomes.
MUDAR v. tr. 1. Dar o tomar otro ser o naturaleza, otro estado, forma o lugar. 2. Dejar algo que antes se tenía y tomar en su lugar otra cosa. 3. Variar, cambiar. Mudar de dictamen, de criterio. 4. Dejar la casa que se habita y pasar a vivir en otra. Mudarse de casa. 5. Ponerse otra ropa o vestido, dejando el que antes se llevaba puesto. Mudar de vestido. 6. Dicho de un animal, renovar periódicamente la piel, las plumas. El pájaro ha mudado las plumas.
Ajuntament de Barcelona
Institut de Cultura
Museu Tèxtil i d'Indumentària
EL MUSEU ES MUDA
EL MUSEU TÈXTIL I D'INDUMENTÀRIA
MUDA DE PELL
Rehabilitem l'edifici per recuperar l'esplèndid Palau del Marquès de Llió al carrer de Montcada.
Millorem la qualitat de la visita dels espais expositius (il·luminació i aire condicionat) i de l'ús del centre de documentació (horaris més llargs, espais més amplis, nous sistemes de consulta).
Transformem els espais. L'estiu de 2008, les sales del palau del carrer de Montcada seran la nova seu d'activitats del futur centre del disseny.
MUDA DE LLOC
Reubiquem la col·lecció del Museu i l'exposició permanent al Palau Reial de Pedralbes, en unes millors condicions físiques de conservació.
MUDA DE CONTINGUT
Actualitzem el discurs de la col·lecció permanent del Museu, que es farà des de la perspectiva del cos humà, a partir del juny de 2008.
En l'àmbit internacional, aquesta és la primera vegada que un museu del vestit presenta la seva col·lecció permanent a partir de la morfologia del cos.
EL MUSEO SE MUDA
EL MUSEU TÈXTIL I D'INDUMENTÀRIA
MUDA DE PIEL
Rehabilitamos el edificio para recuperar el espléndido Palacio del Marqués de Llió en la calle Montcada.
Mejoramos la calidad de la visita de los espacios expositivos (iluminación y aire acondicionado) y del uso del centro de documentación (horarios más prolongados, espacios más amplios, nuevos sistemas de consulta).
Transformamos sus espacios. En verano de 2008, las salas del palacio de la calle Montcada serán la nueva sede de actividades del futuro centro del diseño.
MUDA DE LUGAR
Reubicamos la colección del Museo y la exposición permanente en el Palacio Real de Pedralbes, en unas mejores condiciones físicas de conservación.
MUDA DE CONTENIDO
Actualizamos el discurso de la colección permanente del Museo, que partirá de la perspectiva del cuerpo humano, a partir de junio de 2008.
En el ámbito internacional, esta es la primera vez que un museo del vestido presenta su colección permanente a partir de la morfología del cuerpo.
WWW.MUSEUTEXTIL.BCN.CAT
* Durant el període de remodelació del Museu, la Biblioteca mantindrà els seus serveis habituals.
* Durante el período de remodelación del Museo, la Biblioteca mantendrá sus horarios habituales.

BNE
空調器入れ
BNE
BNE

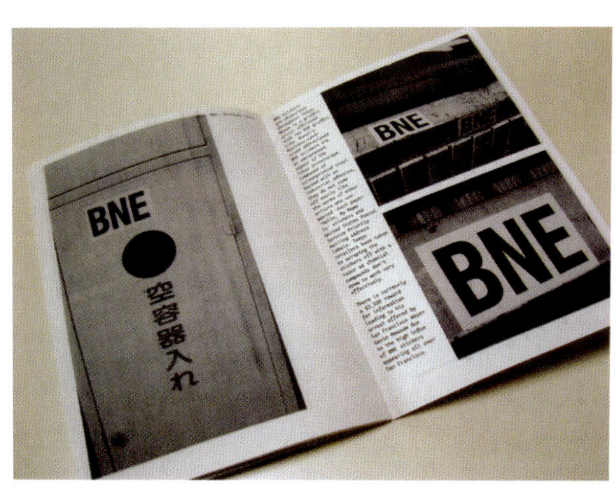

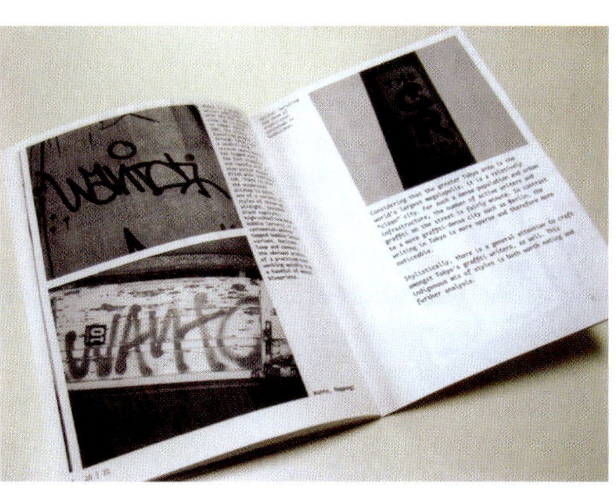

A playful typographic composition was created using woodblock letters and then printed on the back of various stationery items. It was adopted as the company's identity even though it doesn't actually give its name – the latter is simply printed in a clean sans serif along with the necessary address information.

An invitation to a trade show is mailed out in a black rubber vacuum-packed bag, which highlights the strange three-dimensional shape within. When the bag is opened, the invite itself is found to consist of a bright yellow bootlace wrapped around a typographic card. Black and yellow are the colours most commonly associated with Dr Martens, especially the bright yellow lace. The bold sans-serif type is bound by the lace.

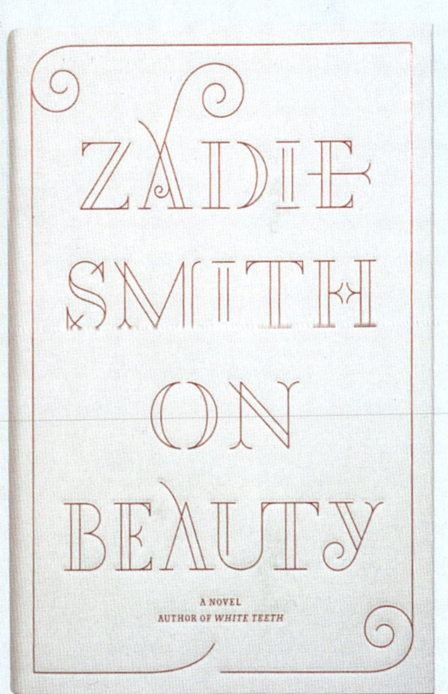

g N

ABCDEFGHIJKLMNOPQRSTUVWXY
ZÀÁÂÃÄÅĀĄÆĆĈČÇĎĐÈÉÊĔĚ
ĘĒĜĞĠĢĦĤÌÍÎÏĨĬİĲĴĶĹĽĿŁÑŃŅŇŊÒ
ÓÔÕÖØŌŎŐŒŔŘŚŜŞŠȘŢŤŦÙÚÛÜ
ŨŪŬŮŰŲŴÝŶŸŹŻŽÞ

abcdefghijklmnopqrstuvwxyzßàáâ
äåāąæćĉčçďđèéêĕěęēĝğġģ
ġĥħìíîïĩĭıĳĵķĺľŀłñńņňŋòóôõöøōŏő
œþŕřśŝşšșţťŧùúûüũūŭůűųŵýÿŷźżž

ABCDEFGHIJKLMNOPQRSTUVWXYZÀ
ÁÂÃÄÅĀĄÆĆĈČČĎĐÈÉÊĔĚĘĒĜĞ
ĢĞĦĤÌÍÎÏĨĬİĲĴĶĹĽĿŁÑŃŅŇŊÒÓÔÕÖ
ØŌŎŐŒŔŘŚŜŞŠȘŤŢŦÙÚÛÜŨŪŬŮŲŰ
ŴÝŶŸŹŻŽÞ

 fffiflffffiflfbfkfhfjftffj

123456789 0123456789 0123456789

!"#$%&'()*+,-./:;<=>?@{|}[\]...¿¡¶•~†‡
€¢£¤¥f§«‹›»"„"‚'"_- ™©®º123ªº¼½¾¡
^‰/¬≠≤≥±×÷µ

#%&*!

flåsk±

READ

Yêăħ;

{©}

ZZZZ
ZZZ
eee
eeee

67 f

7 Pt | Line Gap 10.5 Pt

¶ Man and horse were littered with leaves and dusted with yellow pollen, for the open was ventured no more than was compulsory. They kept to the brush and trees, and invariably the man halted and peered out before crossing a dry glade or naked stretch of upland pasturage. He worked always to the north, though his way was devious, and it was from the north that he seemed most to apprehend that for which he was looking. He was no coward, but his courage was only that of the average civilized man, and he was looking to live, not die. Up a small hillside he followed a cowpath through such dense scrub that he was forced ...

IMPORTANT: For best results from this material allow it to reach equilibrium with room temperature before using this Font.

≤ 110 ≥ WORK ORDER / TRIMATIC™ BBP
RELATIVE HUMIDITY ‹ 1.2.3 ›
AT 20°C: 50% ± 5%
FORMAT TYPE D ≤+≥ K: 58 × 34

EARTH-TRACKS
ΓΗ-ΙΧΝΗ
19 12 09 > 28 2 10
MUSEUM
OF PHOTOGRAPHY
THESSALONIKI
ΜΟΥΣΕΙΟ
ΦΩΤΟΓΡΑΦΙΑΣ
ΘΕΣΣΑΛΟΝΙΚΗΣ
TRACKS
EARTH
ΓΗ
ΙΧΝΗ
PRIX PICTET
TO BHMA

TRACKS
EARTH
ΓΗ
ΙΧΝΗ

EARTH
ΓΗ

Produced to accompany an exhibition of Wim Crouwel's posters from the late 1950s to the early 1970s, this catalogue uses a font originally designed by Crouwel for a show of work by American Pop artist Claes Oldenburg but that has been digitally redrawn by The Foundry. The face echoes Oldenburg's 'soft sculptures', which converted everyday objects into larger-than-life soft PVC versions of themselves.

JULIET
&
ROMEO

description
A promotional poster demonstrating
the full range of typefaces in the
Shire Types family. It also gives
information on the concepts behind
their creation.

dimensions
410 x 544 mm
16¼ x 21⅜ in

W:THEM
PLEASE
FEEL ABSO-
LUTELY
FREE TO
CONTACT
ME...
MEGABEN [AT] PLOKK.DE

BO
MB
IA

Kunstbücher,
Fotobücher,
Ausstellungs-
kataloge
Art and Photo-
graphy, Exhibi-
tion Catalogues
5

REBELLE
CEO GRAD JE BIOSKOP!
9 DANA
15 LOKACIJA
125 FILMOVA
FILM & MEDIA FEST
CINEMA CITY
06-14. jun, 2009. Novi Sad
cinemacity.org

everybody Loves
TYPE
typefaces

Typeface is a coordinated set of glyphs
designed with stylistic unity. A typeface
usually comprises an alphabet of letters,
numerals, and punctuation marks; it may also
include ideograms and symbols, or consist
entirely of them, for example, mathematical
or map-making symbols. The term typeface is
typically conflated with font, which had
distinct meanings before the advent of desktop
publishing. These terms are now effectively
synonymous when discussing digital typography.
One still valid distinction between font and
typeface is that a font may designate a specific
member of a type family such as roman, bold or
italic type, possibly in a particular size, while
typeface designates a visual appearance
possibly of a related set of

SAUSAGES
AND ERROR MESSAG-
ES
SUNDAY GRILL + VISTA WORK-
SHOP BRING SAUSAGES AND LAP-
TOPS: WE'LL MIX THEM UP
SOMEHOW
OPEN
WORKSHOP
THIS
19.12.08 SUNDAY
20:00
SUNDAY GRILL +
VISTA WORKSHOP
BRING SAUSAGES AND
LAPTOPS: WE'LL MIX
THEM UP SOMEHOW
SUNDAY   19.12.08
              20:00

ROBODONUTS VS. MECH-
KEBABS
ROBOTIC COOKING WORK-
SHOP BRING RECIPES AND
OTHER NECESSATIES
OPEN
WORKSHOP
THIS
19.12.08 SUNDAY
20:00
SUNDAY GRILL +
VISTA WORKSHOP
BRING SAUSAGES AND
LAPTOPS: WE'LL MIX
THEM UP SOMEHOW
SUNDAY   19.12.08
              20:00

collage

abcdefghijklmn
opqrstuvwxyz
1234567890 (.,:;!?&)

The A5 series is intended as a growing archive on the history of graphic design. Each volume introduces outstanding personalities and important themes from the history of international graphic design, with numerous illustrations, essays and interviews. A5 is a co-operative project between the labor visuell in the design department at the Fachhochschule Düsseldorf and Lars Müller Publishers. The first three books, which have been published October 2009, are about Hans Hillmann, one of Germanys most influential graphic designers, a forgotten series of record-covers designed by legendary artists like Max Bill or Karl Gerstner and the legendary white paperback of 'dtv' with fascinating cover designs by swiss designer Celestino Piatti. Each volume is in German and English language.

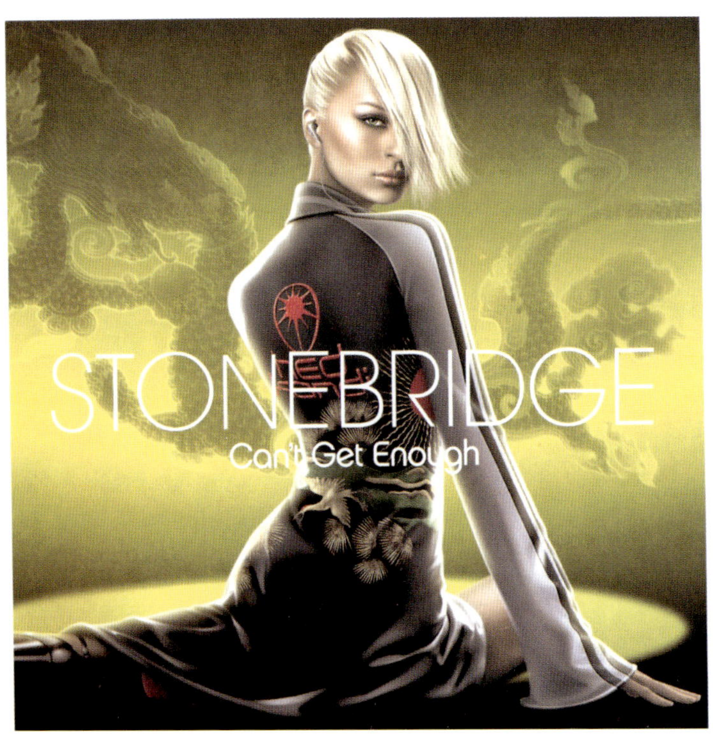

"I love to experiment with new forms and styles of expression,
and my greatest passion is to embody the feminine soul in my illustrations."

A spread from the Columbus Society
of Communicating Arts Creative
Best Catalog, which showcases the
winners of the Columbus Society
annual competition. By reassembling
the winning entries into a collage,
which were then used as graphics
throughout the book, the 1996
catalog became a representation
of the exhibition while still retaining
its own identity.

artwork title
Striplight

typeface
Striplight

designer
Craig Yamey

design company
YAM

photographer
Craig Yamey

country of origin
UK

Slobodan Jelesijević   Slobodan Jelesijević

description
A still from a linear, interactive
narrative exploring the
subconscious of an individual
with extraordinary powers.

For four years Non-Format art-directed this long-established music monthly. A clean simple grid system and sans-serif body-copy style sheets were first established. The designers then created new headline fonts for nearly every number, while other issues used an existing font and welded distressed elements onto it. Over the four-year period the magazine retained an identity all its own through its creative use of experimental headline fonts, white space and excellent photography.

IMPORTANT: For best results from this material allow it to reach equilibrium with room temperature before using this Font.

⧼110⧽ WORK ORDER / TRIMATIC™ BBP

DK RELATIVE HUMIDITY ⟨1.2.3⟩
AT 20°C: 50% ± 5%
FORMAT TYPE D⧼+⧽K: 58 x 34

República 1053
4700   San Fernando del Valle
de Catamarca
03833 431 352 | 435 968
info@yucuco.com
www.yucuco.com
FINCA YUCUCO

Certificate
$5.00 OFF
Your next visit
niccolo SALON & spa

OUT
NOT
100

The designers created a complete graphic identity system
for Stedelijk Museum CS, the temporary exhibition space of
the Amsterdam museum of modern art. This included
logotype, signage system, bimonthly magazine and
exhibition posters. The identity renders the letters (set in
a bold sans-serif font) in a series of diagonal lines in cyan
and red. This diagonal line is used throughout the other
applications together with the clean sans font printed in
cyan and red.

A B C D E F G H I J K L M N Ñ
O P Q R S T U V W X Y Z

a b c d e f g h i j k l m n ñ o p
q r s t u v w x y z

0 1 2 3 4 5 6 7 8 9

Á É Í Ó Ú Ä Ë Ï Ö Ü À È Ì Ò Ù
á é í ó ú ä ë ï ö ü à è ì ò ù

! % [=] { ? } – [ ] . , : ;

[({^`¨¯´ ˜°ˇ˝~*.,;…"'„‖†‡^‹'""•_–—˜})]
àáâãäåæçèéêëìíîïòóôõøùúûüýÿ
ÀÁÂÃÄÅÆÇÈÉÊËÌÍÎÏÑÒÓÔÕÖØÙÚÛÜÝŸ
ªº123™#©$¢ß€£¥þ ¤µ¡!¿?/\‰%‹›«»#&€§@¶-+±÷=¬<>

デザインの宿題。
Low Carbon
Life-design Award
2009
CO²-25%

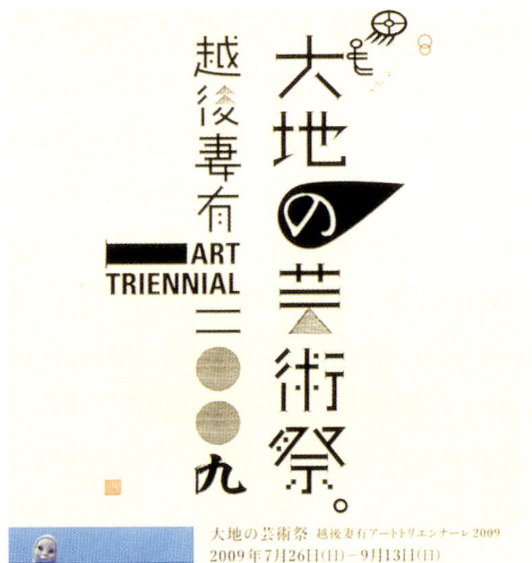
大地の芸術祭。
越後妻有
ART TRIENNIAL
二〇〇九
大地の芸術祭 越後妻有アートトリエンナーレ 2009
2009年7月26日(日)-9月13日(日)
http://www.echigo-tsumari.jp

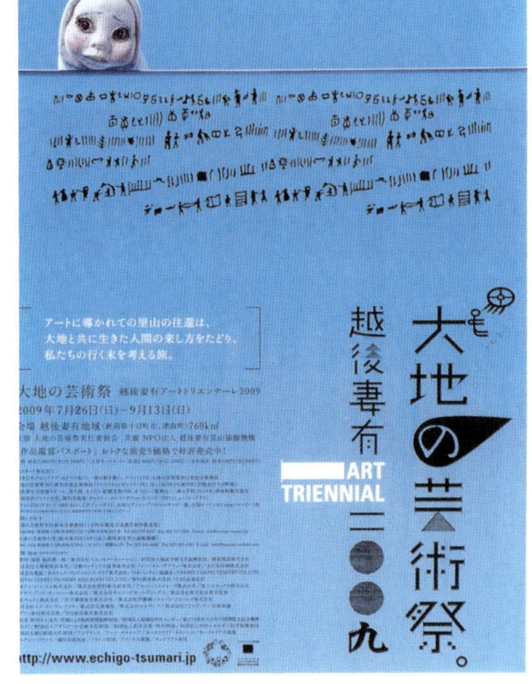
大地の芸術祭。
越後妻有
ART TRIENNIAL
二〇〇九
大地の芸術祭 越後妻有アートトリエンナーレ 2009
2009年7月26日(日)-9月13日(日)
http://www.echigo-tsumari.jp

MUSHROOM

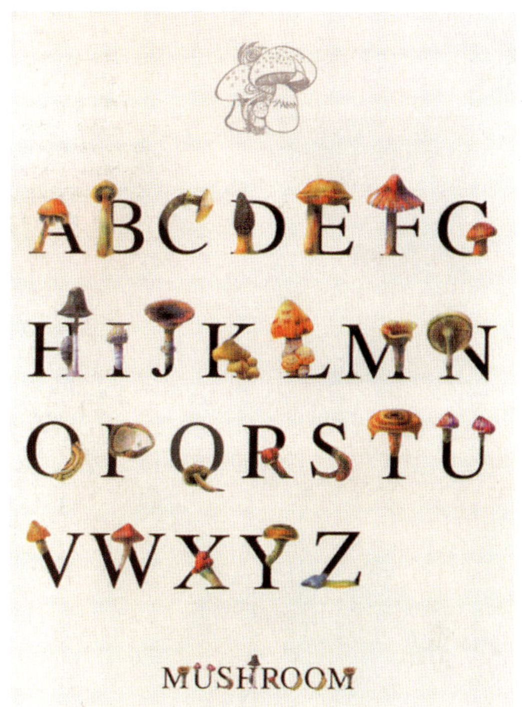
ABCDEFG
HIJKLMN
OPQRSTU
VWXYZ
MUSHROOM

比較
日本
文化
研究
Journal of
Comparative Studies
in Japanese Culture
第 9 号
2005年11月
比較日本文化研究会
【小特集】
「子ども」とはなにか
子どもというテーマ　堀田穣
中世の子どもの労働　斉藤研一
──なぜ子どもが売買されたのか
地理学における子ども研究　大西宏治
台湾博覧会　山路勝彦
──植民地は今花盛り
中世の囃子物芸能　青盛透
──囃子の民俗性をめぐって
近世女帝像の形成　野村玄

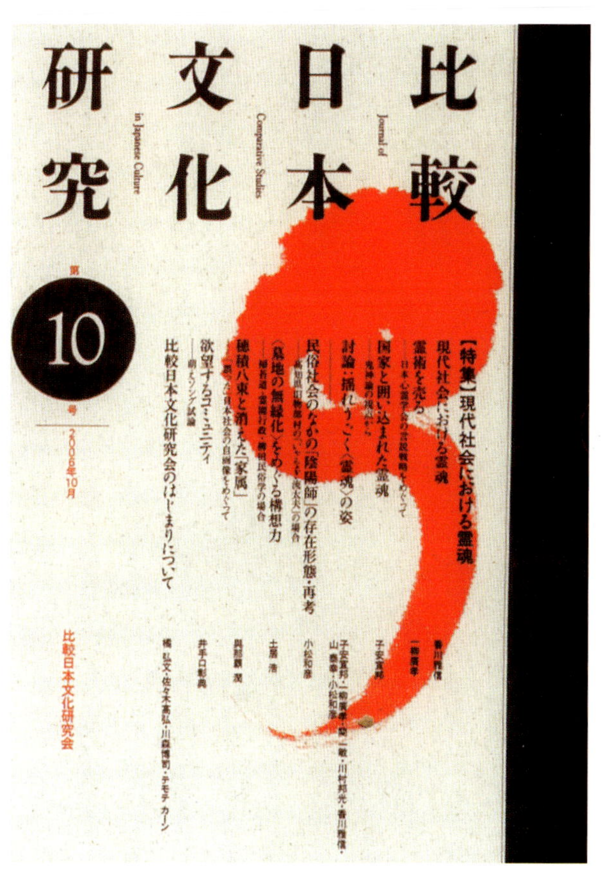
比較
日本
文化
研究
Journal of
Comparative Studies
in Japanese Culture
第 10 号
2006年10月
比較日本文化研究会
【特集】現代社会における霊魂
現代社会における霊魂
霊術を売る　一柳廣孝
──日本心霊学会の言説戦略をめぐって
国家と囲い込まれた霊魂　子安宣邦
討論：揺れうごく〈霊魂〉の姿
──鬼神論の現在から
民俗社会のなかの「陰陽師」の存在形態・再考　小松和彦
〈墓地の無縁化〉をめぐる構想力　土居浩
──柳田・常民行政・柳田民俗学の場合
穂積八束と消えた「家属」
──近代日本社会の自画像をめぐって
欲望するコミュニティ
──ボランティア試論
比較日本文化研究会のはじまりについて
比較日本文化研究会

REVNE
ROOM

cinema
portugal
2007

cinema
magazine
portugal
2006

cinema
portugal
2006
contents

UM POUCO MAI
PEQUENO
QUE
INDIANA

artwork title
Hønsenfest

typeface
bf_BiaBia

designer
Rolf Zaremba

design company
HP Albrecht Advertising Agency

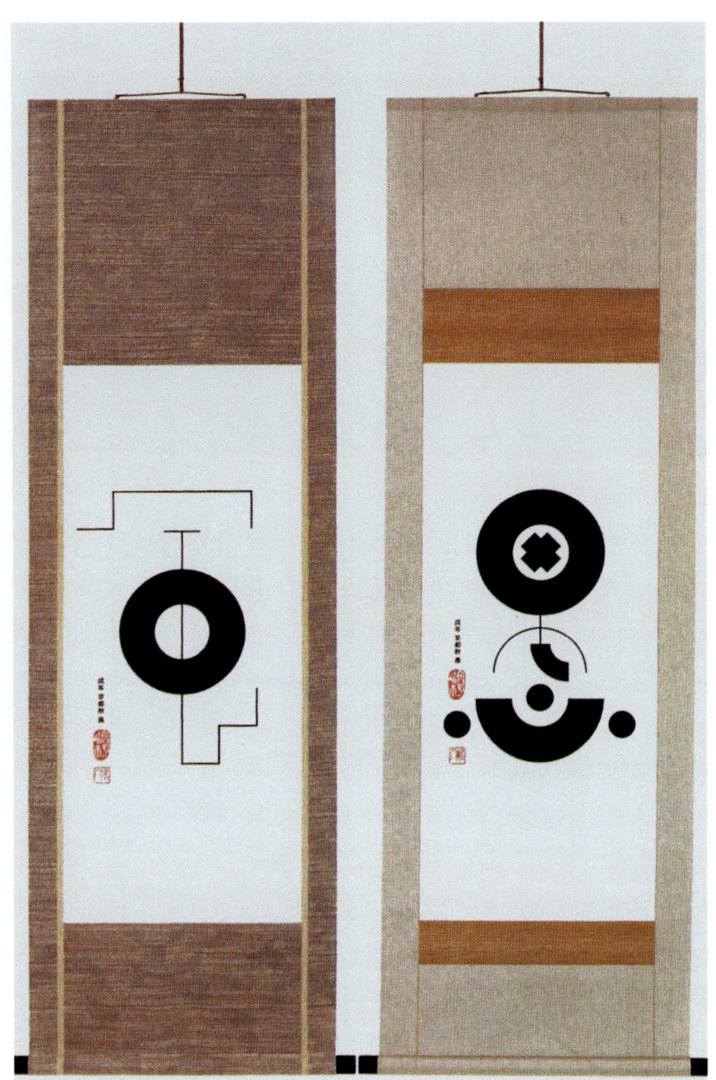

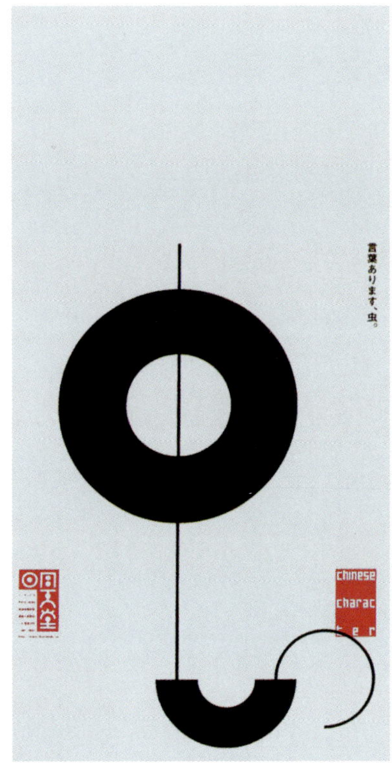

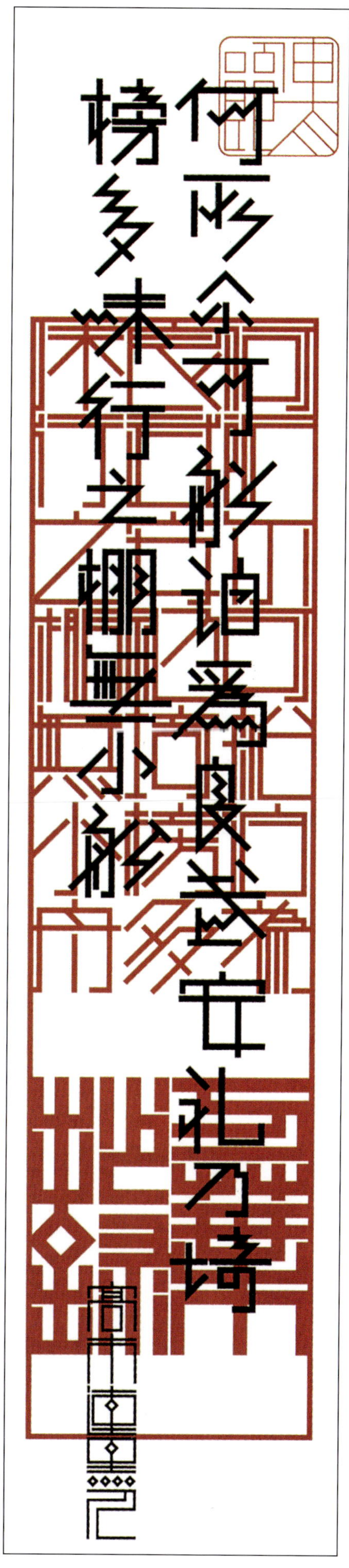

2002 15. - 29. MAJ
DANSEFORESTILLING
SHE LOVES YOU
KUN 13 FORESTILLINGER THE BEATLES
KÆRLIGHEDSSANGE FORTOLKET I DANS
WWW. AVENY-T.DK PETER SCHAUFUSS
BILLETBESTILLING 7020 1031
BILLETNET 7015 6565
AVENY-T

INSTRUKTION
PETER REICHHARDT
SCENOGRAFI
CAMILLA BJØRNVAD
MUSIK
SØREN SIEGUMFELDT
MED
TAMMI OST
MORTEN EISNER
LAURA BRO
PETER MYGIND
GUNNVA ZACHARIASEN
CARSTEN BJØRNLUND
RASMUS SORINUS
ANDERS LUNDEN KJELDSEN/
OSCAR RANCH GAARDBO
8. MAJ
12. MARTS
SOYA/BØRNEDAL'S
EN GÆST
AVENY-T

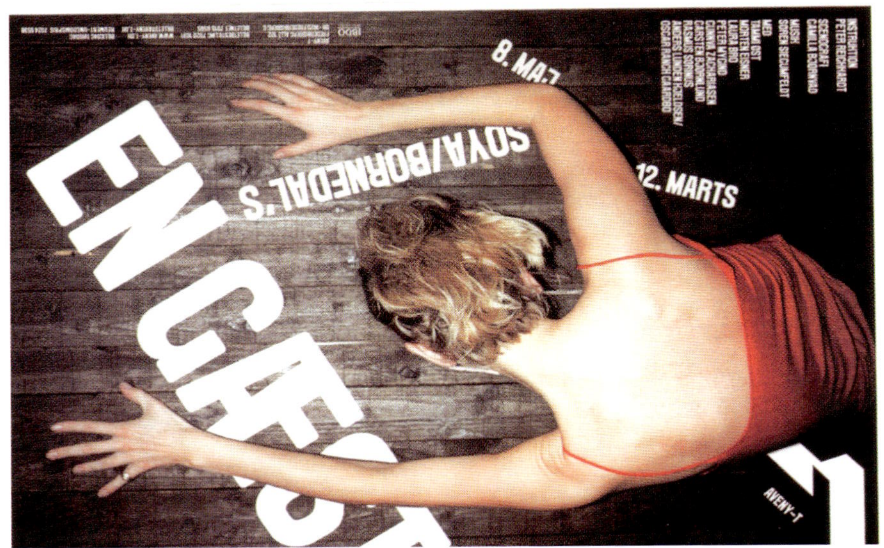

typographic layouts. For every section we used
different paper. In addition the book contains three
own designed typefaces. We used a lot of different
Material (paper) and alltogether it comes with
three typographic posters in a paper case finished
with varnish. The Poster representing the various
Typfaces which were designed for the book. A
die cut is used for the navigation in the book and
is presented in the first pages. It has no page
numbers and for every section the die cut has a
specific designed symbol. After every section there
is a divider page which is binded in the book as
chinese block binding. These pages are working as
title pages for every section.

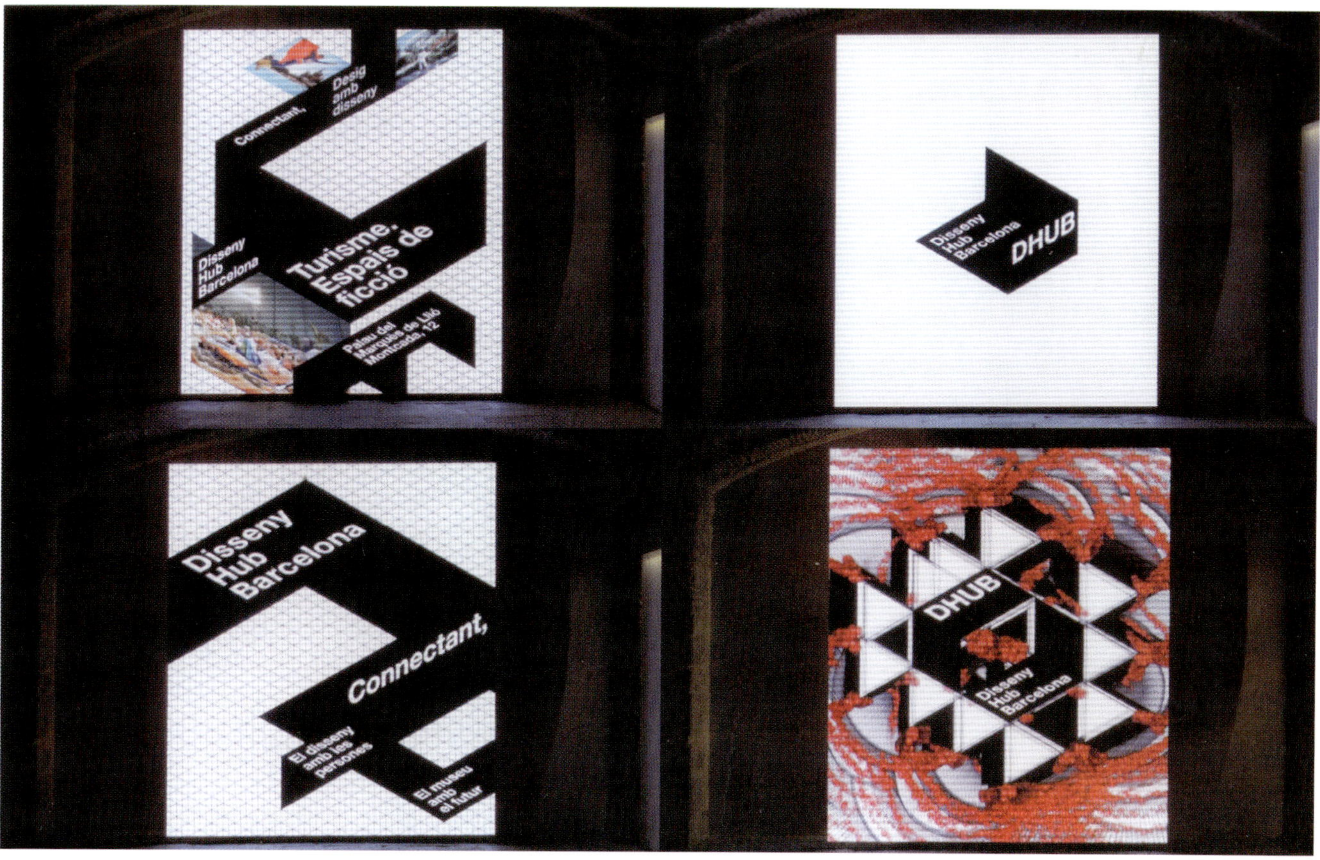

LOVE AND
Obstacles
ALEKSANDAR HEMON
STORIES
AUTHOR OF THE LAZARUS PROJECT

PLAS
TIQUE
EXPLOSIVE FASHION

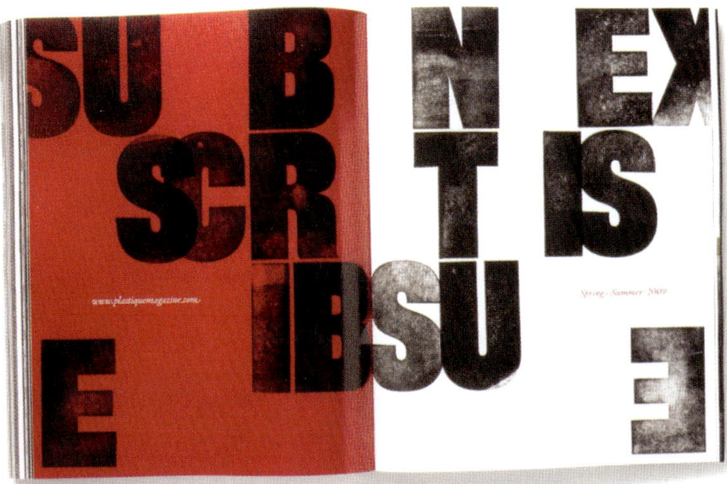

SU BN EX
SCR T IS
E IB SU
E E
www.plastiquemagazine.com
Spring - Summer 2009

Environmental Writing Since Thoreau
AMERICAN
EARTH
FOREWORD BY AL GORE
EDITED BY BILL McKIBBEN

Little
Bee
a novel
CHRIS CLEAVE

artwork title
TypeFace#5

typeface
BD Doomed SquareUp

designer
MBrunner

design company
büro destruct

country of origin
Switzerland

description
as on page 31

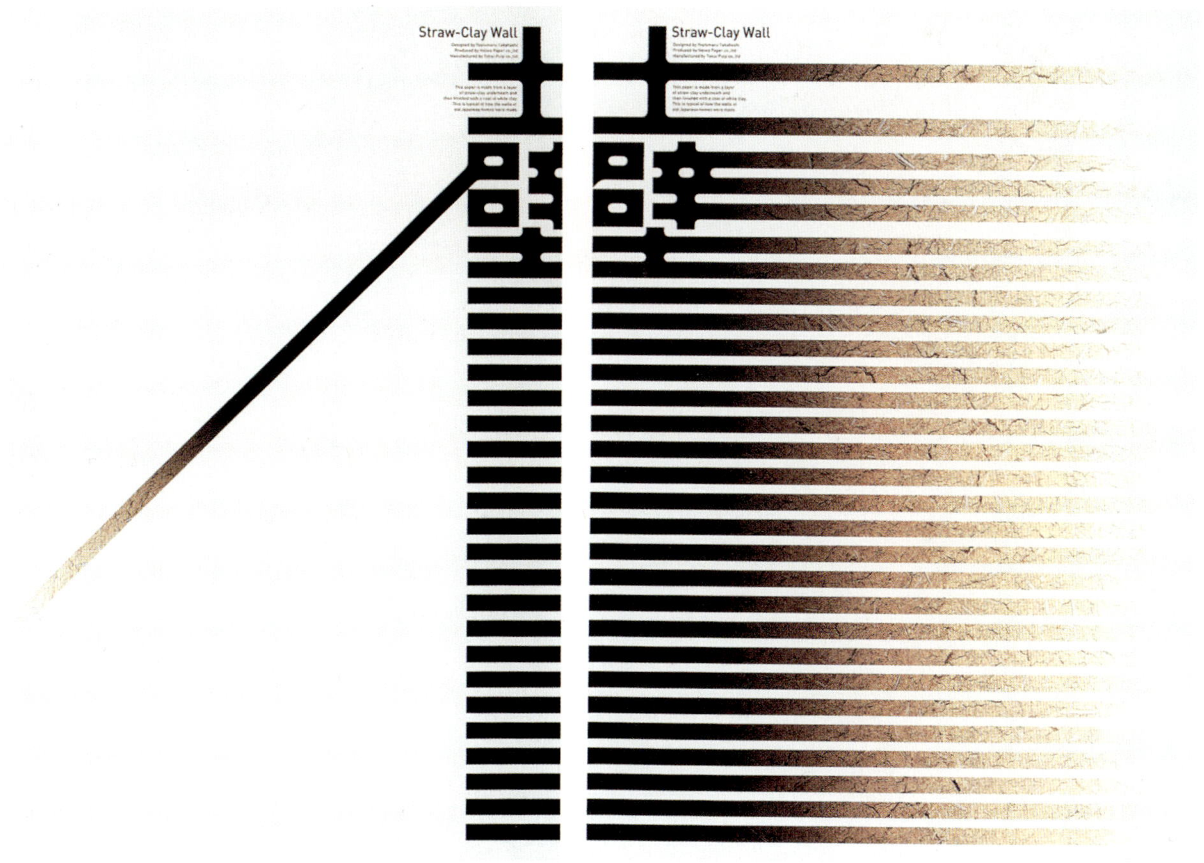

artwork title
Rabi!

typeface
Shark

designer
Atsushi Aoki

design company
Atsushi Aoki

illustrator
Atsushi Aoki

VILCEK
PRIZE 2010
JOAN
MASSAGUE
VILCEK
PRIZE
CHRISTO &
JEANNE-CLAUDE

The different disciplines available at the Graduate School of Arts, Culture and Environment are constructed out of three-dimensional letterforms for this poster for the University of Edinburgh. An aerial shot of the city, printed in green and black, forms the background against which the thin white grid lines used for the construction of the headline typography are visible. A different colour is employed for each subject, which helps legibility since the letters otherwise become interwoven over the poster.

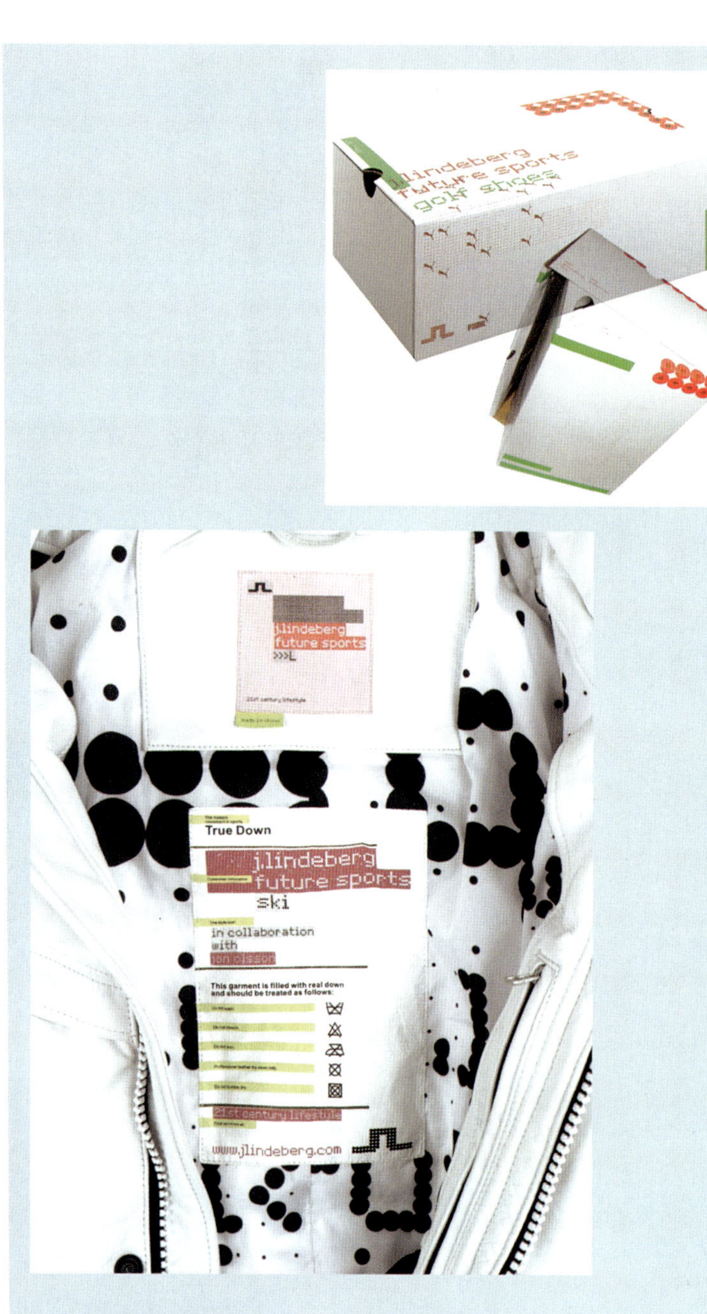

AIRPLOT!
STOP THE EXPANSION
AIRPLOT!
ABCDEFGHIJKLMNOPQRSTUVWXYZ
ABCDEFGHIJKLMNOPQRSTUVWXYZ
ABCDEFGHIJKLMNOPQRSTUVWXYZ
ABCDEFGHIJKLMNOPQRSTUVWXYZ
ABCDEFGHIJKLMNOPQRSTUVWXYZ
ABCDEFGHIJKLMNOPQRSTUVWXYZ

JOIN THE PLOT!

GREENPEACE PRESENTS
AIRPLOT!
NO THIRD RUNWAY

NØRREBROS
TEATER

mutek iTXS
Festival Internacional de Cultura Digital, Quinto Aniversario
Octubre 23 - 26, 2008 Ciudad de México
GUILLAUME & THE COUTU DUMONTS_CA
PLEIN DU SOLEIL feat CHLOÉ + KRIKOR_FR
LATINSIZER_MX
VINCENT LEMIEUX_CA
CUBENX_MX
ELTONO_FR
SPECTRUM_MX
WIGHNOMY BROTHERS_DE
NICOLAS JAAR_US
BYETONE_DE
ALVA NOTO_DE
JIMMY EDGAR_US
FAX_MX
CRONIC_FR
ORIGINAL HAMSTER vs NEGO MOCAMBIQUE_CA_BR
ANJA SCHNEIDER_DE
MONOPOLAR_DE
SIGNAL_DE
METRIKA_MX
MATHIAS KADEN_DE
DEADBEAT_CA
301 visuals_CA
BN LOCO_MX
FLYING LOTUS_US
DENDRON_MX
TWINE_US
FRANK BRETSCHNEIDER_DE
PFADFINDEREI visuals_DE
ESTRUCTURAS DE LA TARDE_MX
NORTEC feat. TIJUANA SOUND MACHINE_MX
American Apparel
Montréal PlayGround Global

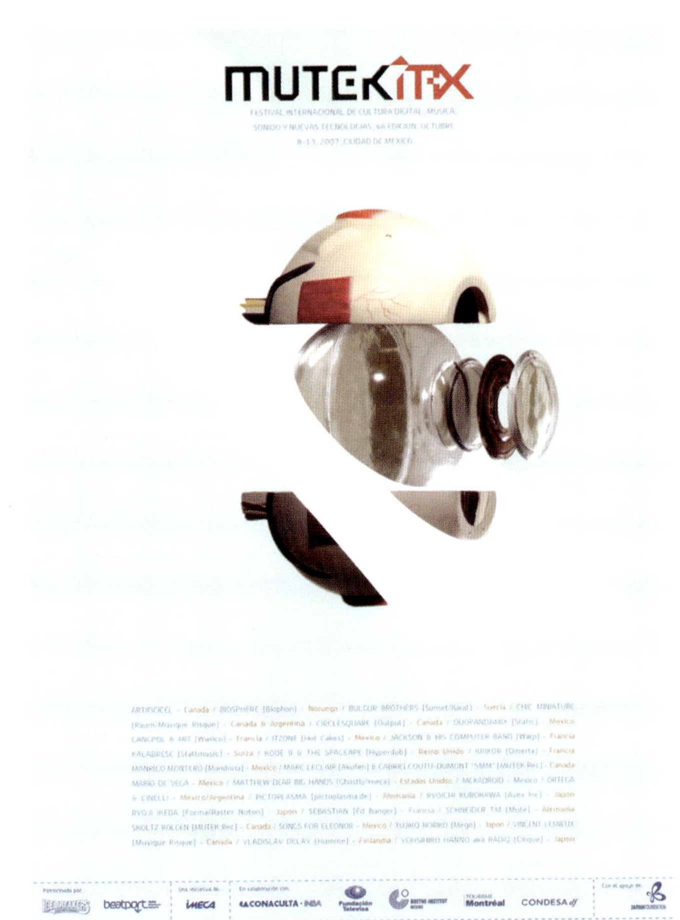

mutek iTX
FESTIVAL INTERNACIONAL DE CULTURA DIGITAL, MÚSICA,
SONIDO Y NUEVAS TECNOLOGÍAS, 4a EDICIÓN, OCTUBRE
8-13, 2007, CIUDAD DE MÉXICO
American Apparel CONACULTA · INBA Montréal CONDESA df

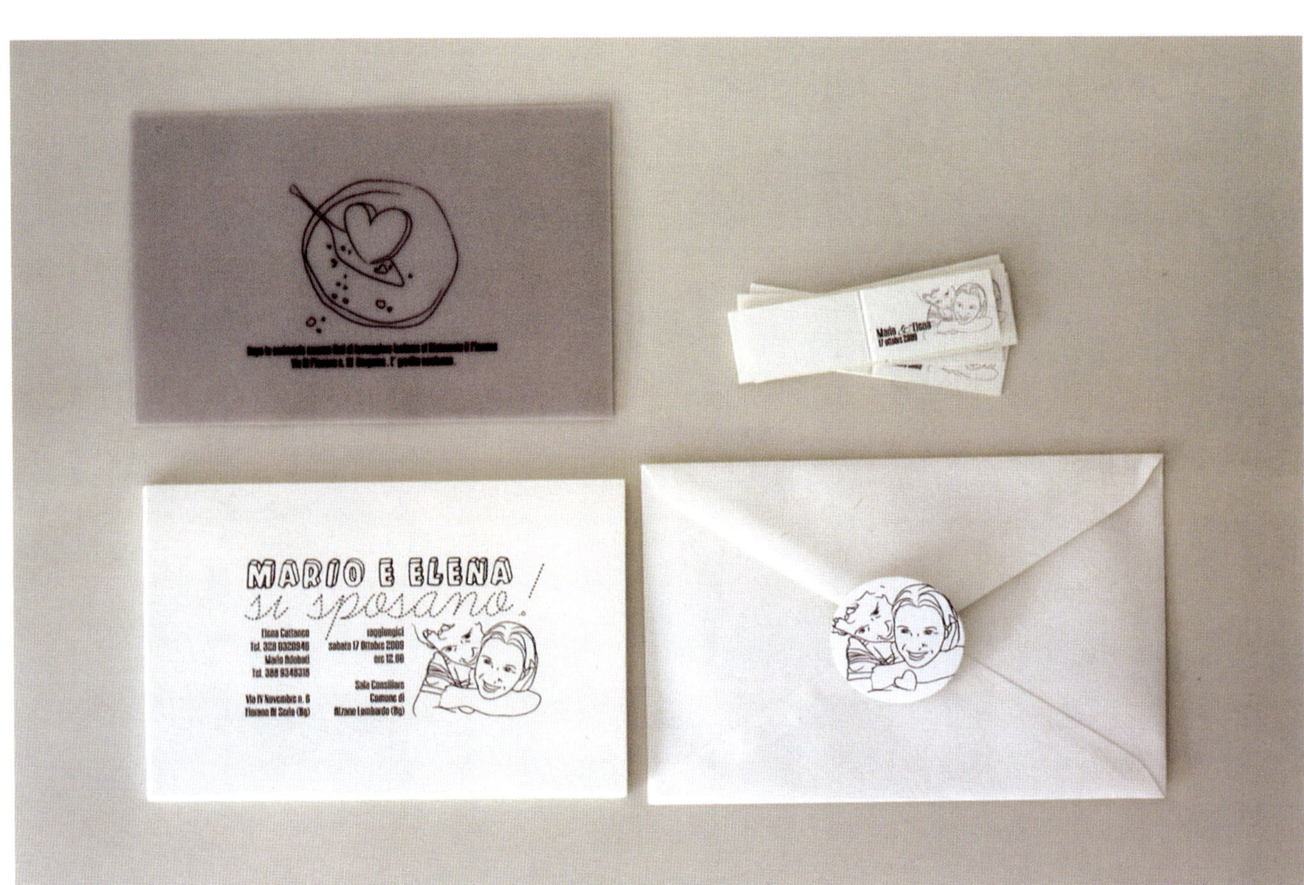

MARIO E ELENA
si sposano!
Elena Cattaneo
Tel. 328 0320946
Mario Adobati
Tel. 389 9349318
Via IV Novembre n. 8
Fiorano Al Serio (Bg)
sabato 17 Ottobre 2009
ore 12.00
Sala Consiliare
Comune di
Alzano Lombardo (Bg)

Harmonie intérieure
Decorative Workshop
www.harmonie-interieure.com
Harmonie intérieure
Collection Urban
Harmonie
atelier
Harmonie intérieure
atelier de décoration
Harmonie intérieure
Collection Organic
Harmonie intérieure
Collection Organic
Édition revendeurs
Reseller edition

Typeworkshop.com originates from workshops given by
Underware, a graphic design studio based in Holland and
Finland that specializes in designing and producing
typefaces. The workshops help the participants to
understand typography more fully in a physical sense.
'Movable Type' used cardboard boxes as its construction
material. 'Manual Pixelism' created pixel-based fonts from
repeated modules, such as disposable plastic drinking
cups, supermarket shopping trolleys and paperback books.
'Let It Run', meanwhile, was constructed as the world's
biggest type-domino.

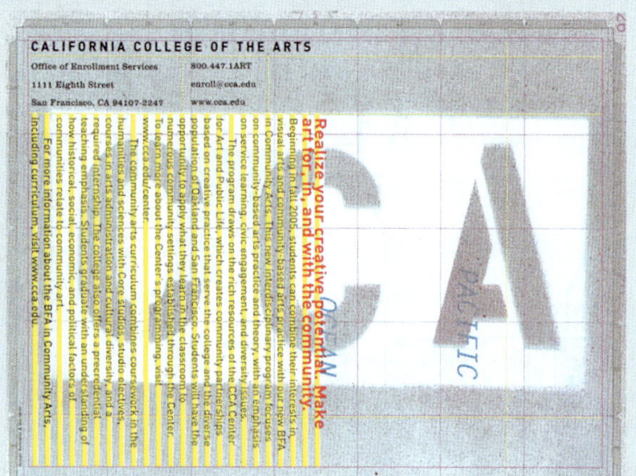

Sinn—
S   h
Sinn—
Sisam
—outh
1935—
1975
Singer—
Songwriter
in the
1950s
19
1975
Songwriter
in the
1950s
November 1975.

Sinn—
Sisamouth
1935—1975
was a Famous
and highly
prolific
Cambodian
singer—
songwriter in
the 1950s
to the 1970s.
1960s
Cambodian
music
scene
Sinn
Sisamouth Typeface
and Dingbats 1.0 Sampler
starting at $39
www.tomtor.com

Sinn—
Sisamouth
with

TESS
IS
MORE

TESS

Ending never

CERT: 2010
CHANGE
COURSE
FUTURE
TRENDS

CERT: curious
CERT: events
invites you to join us for an exclusive
presentation by internationally renewed
australian businessman Simon Hayward.
rsvp:

Identity for a London based model agency. Tess represents well established
names such as Naomi Campbell and Erin O'Connor in the UK. The identity
uses several logo variations based on a modular system of art-deco inspired
elements. The same elements are being used for frames (which overlap images
of the models) on various printed applications as well as on the website.

DUDOK ARNHEM 5 JAAR
DUDOK
ARNHEM
5 JAAR

CHRISTIAN
SURFERS
PORTO
CAMP
Abr
25 a 27
weSC

greenlight
English for Business

open

greenlight
English for Business
M. 620 954 472 | info@greenlight.cat
www.greenlight.cat

...MENTRE PARLEM DE NEGOCIS?

greenlight
English for Business
greenlight
English for Business
C/ Azcárate, 41 1r-1a | 08223 Terrassa

QUÈ TAL
SI FEM
UN TE...

Ruth Harrison:
Tiermaschinen

dtv

Ruth Harrison
Tiermaschinen
[Animal Machines]

Wolfgang Kraus:
Der fünfte
Stand

Aufbruch der Intellektuellen
in West und Ost

dtv

Wolfgang Kraus
Der fünfte Stand
[The Fifth Estate]

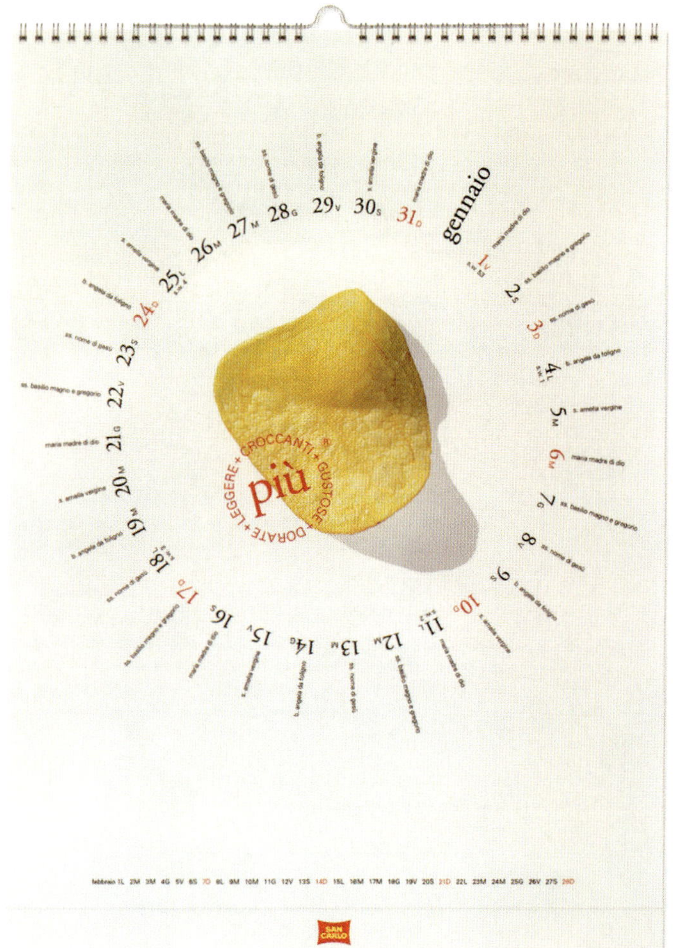

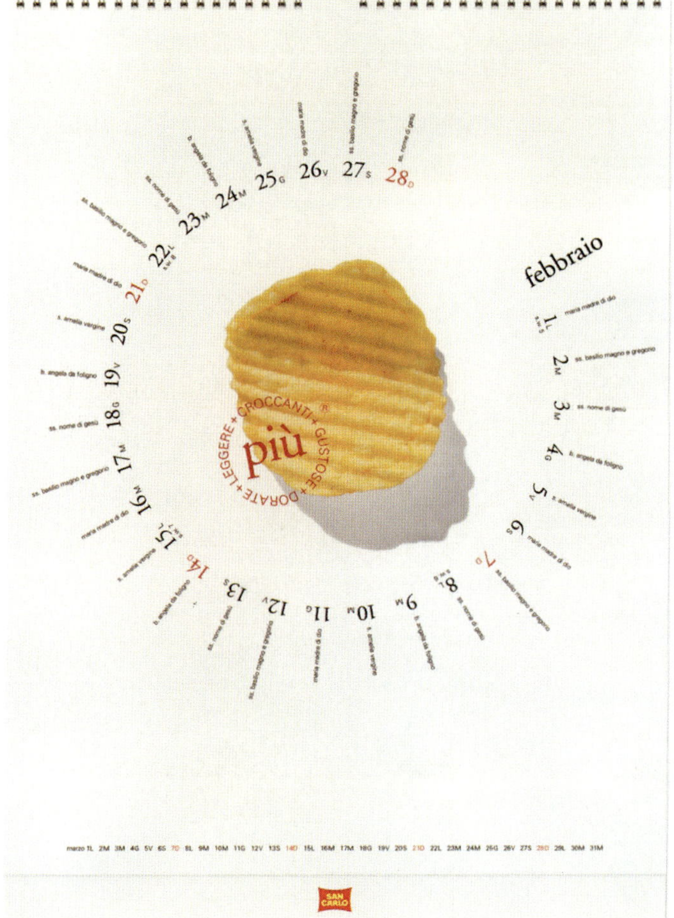

This series of posters for the Forsythe Company combines manipulated overlaid images of the dancers with typographic distortions of the company's name which echo the dynamics of the bodies. The campaign revolves around the idea of non-dimensional space, its abstraction emphasized by the collaging process. The logo appears as a spatial element in different situations as well as being a dynamic, integrated part of the collage itself.

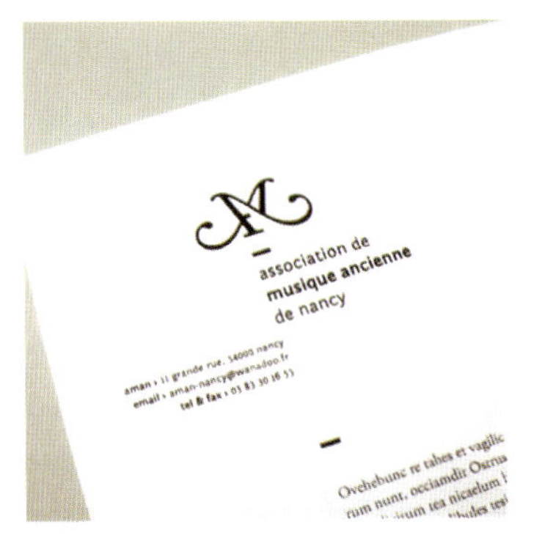

2009 / 2010
SAISON
ANTOINE GUERBER
ROSE
TRES BELE
POLYPHONIES & CHANSONS
DES DAMES TROUVÈRES
ENSEMBLE DIABOLUS IN MUSICA
LUNDI 12 OCTOBRE 2009
20:30 / SALLE POIREL
www.aman-nancy.fr
location : fnac & salle poirel

AKADEMIA
LE PETIT
ORPHEE
FRANÇOISE LASSERRE
LUNDI 9 NOVEMBRE 2009
20:30 / SALLE POIREL
www.aman-nancy.fr
location : fnac & salle poirel
ALICE PIEROT / ALINE ZYLBERAJCH
VIOLON
BAROQUE /
PIANOFORTE
SCHOBERT / MOZART / HAYDN
LUNDI 15 MARS 2010
20:30 / SALLE POIREL
www.aman-nancy.fr
location : fnac & salle poirel

PARAMOUNT

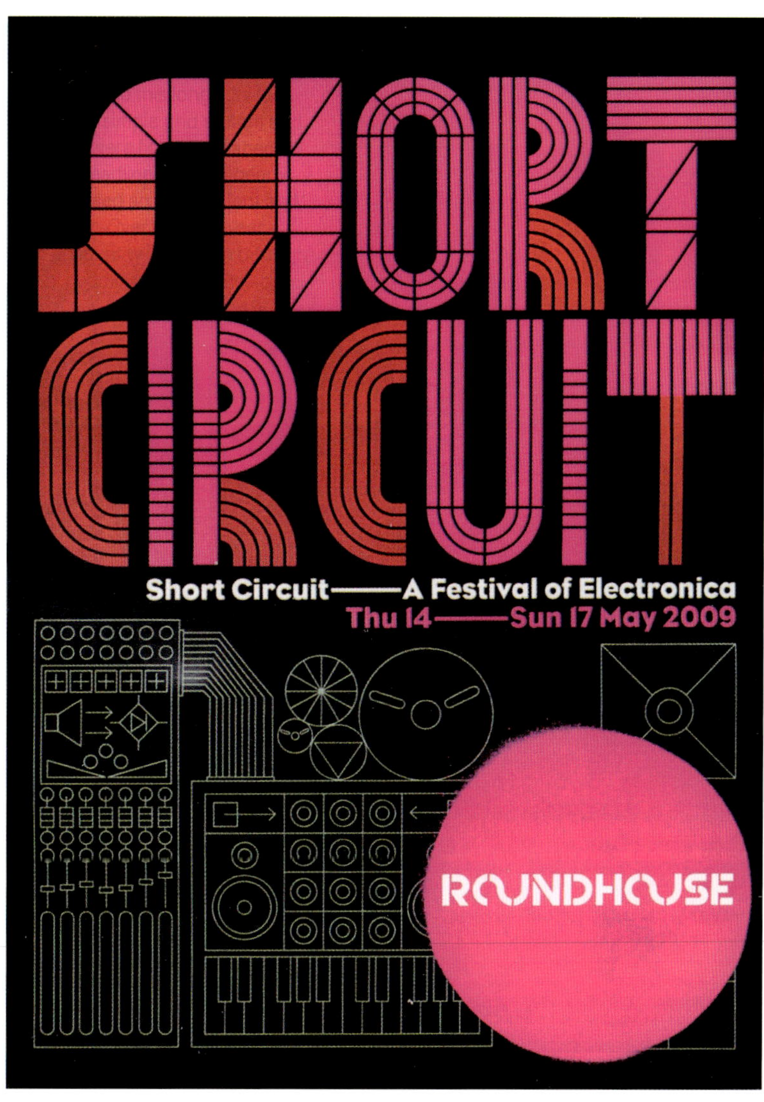

SHORT CIRCUIT
Short Circuit — A Festival of Electronica
Thu 14 — Sun 17 May 2009
ROUNDHOUSE

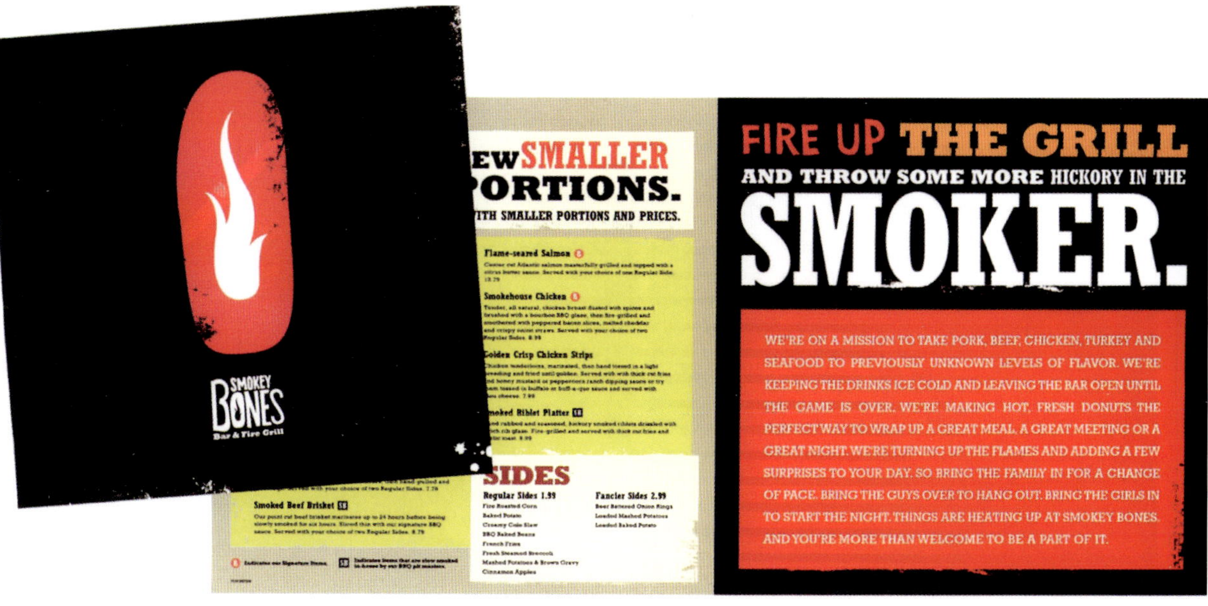

SMOKEY BONES
Bar & Fire Grill
New SMALLER PORTIONS.
WITH SMALLER PORTIONS AND PRICES.
Flame-seared Salmon
Center cut Atlantic salmon masterfully grilled and topped with a citrus butter sauce. Served with your choice of one Regular Side. 12.79
Smokehouse Chicken
Tender, all natural, chicken breast glazed with queso and brushed with a bourbon BBQ glaze, then fire grilled and smothered with peppered bacon slices, melted cheddar and crispy onion straws. Served with your choice of two Regular Sides. 8.99
Golden Crisp Chicken Strips
Chicken tenderloins, marinated, then hand tossed in a light breading and fried until golden. Served with crisp thick cut fries and honey mustard or peppercorn ranch dipping sauce or try them tossed in buffalo or buff-a-que sauce and served with bleu cheese. 7.99
Smoked Riblet Platter
Hand rubbed and seasoned, hickory smoked riblets drizzled with our sweet rib glaze. Fire-grilled and served with thick cut fries and butter toast. 9.99
Smoked Beef Brisket
Our point cut beef brisket marinates up to 24 hours before being slowly smoked for six hours. Sliced thin with our signature BBQ sauce. Served with your choice of two Regular Sides. 8.79

SIDES
Regular Sides 1.99
Fire Roasted Corn
Baked Potato
Creamy Cole Slaw
BBQ Baked Beans
French Fries
Fresh Steamed Broccoli
Mashed Potatoes & Brown Gravy
Cinnamon Apples

Fancier Sides 2.99
Beer Battered Onion Rings
Loaded Mashed Potatoes
Loaded Baked Potato

Indicates our Signature Items.
Indicates items that are slow smoked in-house by our BBQ pit masters.

FIRE UP THE GRILL
AND THROW SOME MORE HICKORY IN THE
SMOKER.
WE'RE ON A MISSION TO TAKE PORK, BEEF, CHICKEN, TURKEY AND SEAFOOD TO PREVIOUSLY UNKNOWN LEVELS OF FLAVOR. WE'RE KEEPING THE DRINKS ICE COLD AND LEAVING THE BAR OPEN UNTIL THE GAME IS OVER. WE'RE MAKING HOT, FRESH DONUTS THE PERFECT WAY TO WRAP UP A GREAT MEAL, A GREAT MEETING OR A GREAT NIGHT. WE'RE TURNING UP THE FLAMES AND ADDING A FEW SURPRISES TO YOUR DAY. SO BRING THE FAMILY IN FOR A CHANGE OF PACE. BRING THE GUYS OVER TO HANG OUT. BRING THE GIRLS IN TO START THE NIGHT. THINGS ARE HEATING UP AT SMOKEY BONES. AND YOU'RE MORE THAN WELCOME TO BE A PART OF IT.

Identity for a London based model agency. Tess represents well established names such as Naomi Campbell and Erin O'Connor in the UK. The identity uses several logo variations based on a modular system of art-deco inspired elements. The same elements are being used for frames (which overlap images of the models) on various printed applications as well as on the website.

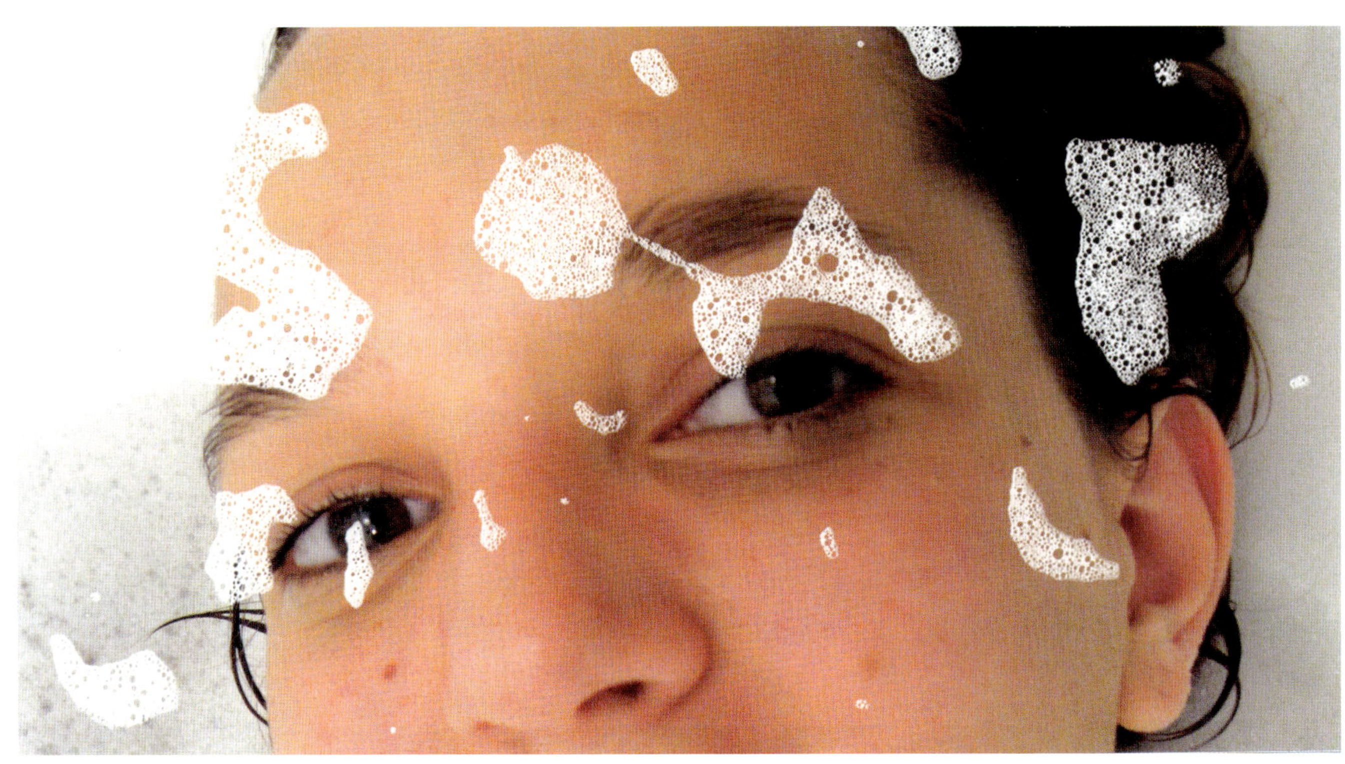

PAUL
paul weeks paul weeks
paul week
WEEK
PAUL
WEEK
weeks PA
paul weeks
PAUL WE
weeks paul weeks paul w
weeks pc

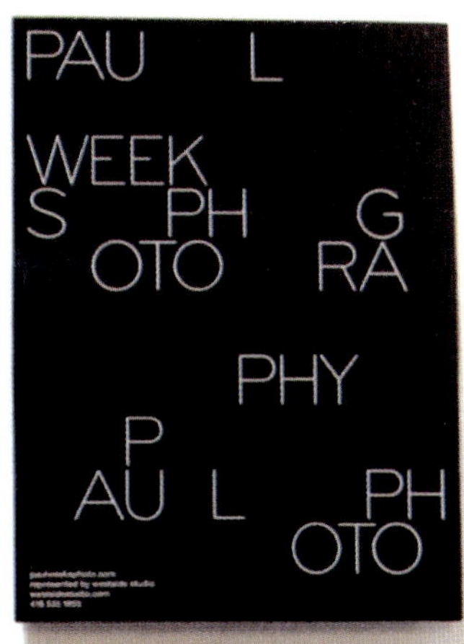

PAU   L
WEEK
S   PH   G
OTO   RA
PHY
P
AU L   PH
OTO
paulweeksphoto.com
represented by westside studio
westsidestudio.com
416 535 1955

paulweeksphoto.com
represented by westside studio
westsidestudio.com
paul
weeks
photo
gr
aphy

PA UL
paulweeksphoto.com
represented by westside studio
westsidestudio.com
416 535 1955

PAUL
WEEKS
Campbell's
SOUP

PAUL
WEEKS

PAUL
WEEKS
EVIDENCE

PAUL
WEEKS
paulweeksphoto.com
represented by westside studio
westsidestudio.com

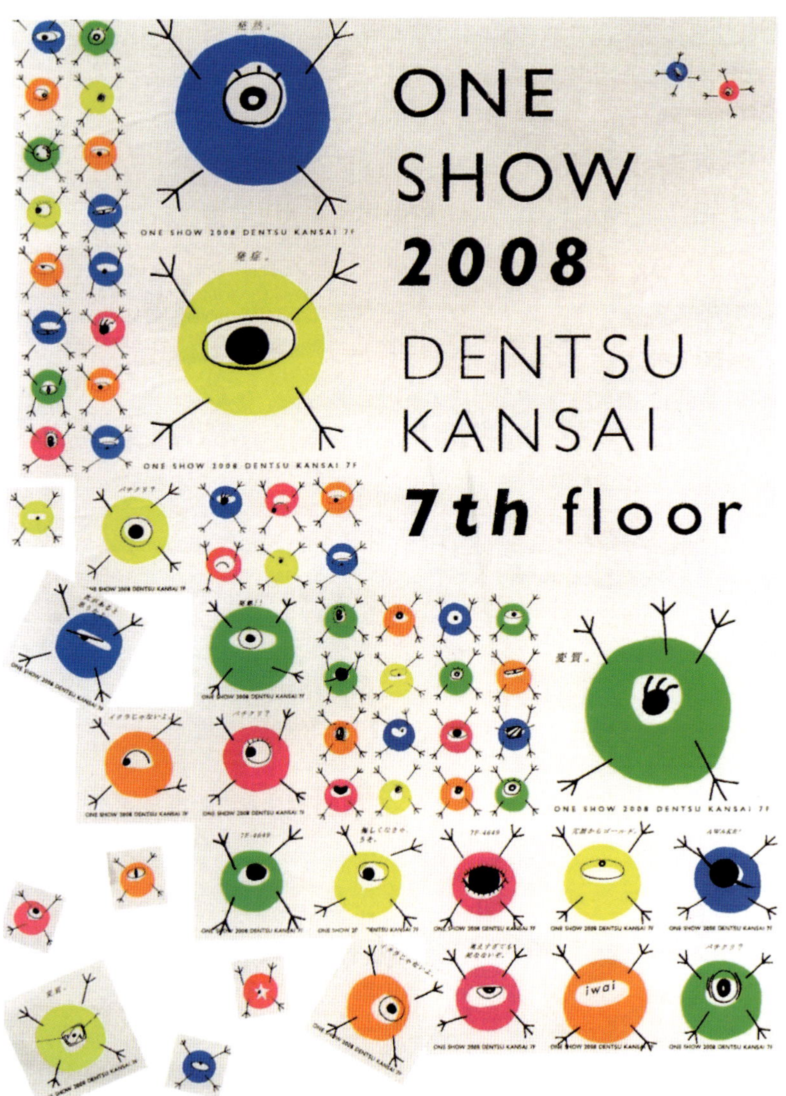

Labor Visuell (visual lab) is a design research unit of Dusseldorfs University of Applied Sciences. In November 2009 it was holding a symposium entitled 'Using design to research design' to mark its 5th anniversary. In addition to special exhibitions and presentations about the various research projects, the event was very much focused on the exchange of ideas and experiences with design researchers. To promote the event we have designed a corporate identity based on five transforming versions of the number five. We have designed flyers, posters and an anniversary booklet. The event itself was designed by a group of students using the elements of our design system.

043_Small graphics: TD. AD. 八木義博 YAGI Yoshihiro/ D. 木村 洋 KIMURA Yo/ C. 筒井晴子 TSUTSUI Haruko/ CL. （株）電通 関西支社 Dentsu Inc. Osaka Branch Office/ PT. Gill Sans Regular, 本明朝 -L

typeface

phliteAudrey

typeface family
phliteAudrey

designer
Hwee Min Loi

foundry/supplier
Hwee Min Loi

country of origin
USA

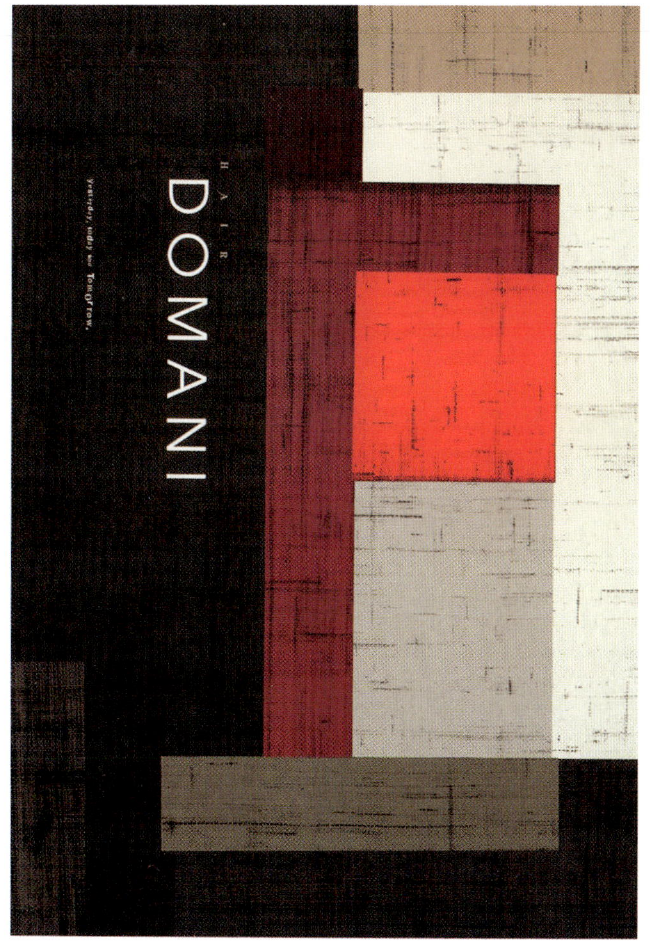

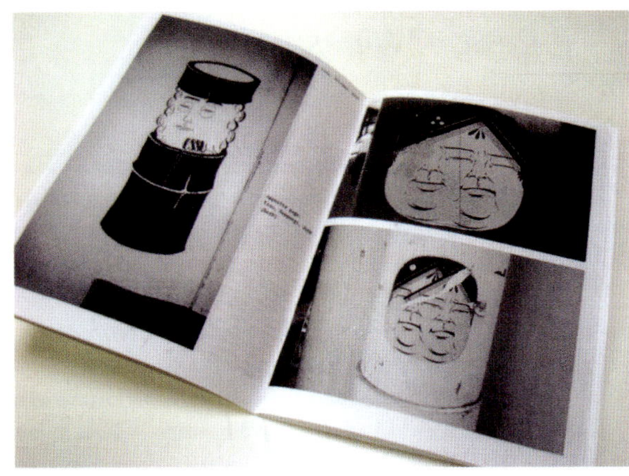
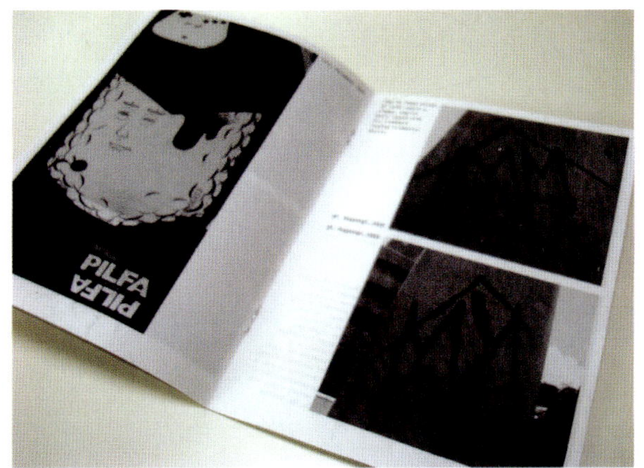

The academy intends this conference to launch a debate on the concrete possibilities for a design that adopts a critical stance vis-à-vis current practice.

The first day will be devoted to the discrepancy between the socio-economic and symbolic reality of the worldwide information and consumer culture, the prospects for a democratisation of the media, and the role of visual producers and theoreticians in this development. The second day is about design that deliberately aims at abolishing the boundaries between everyday and aesthetic experience and will deal with the strategies and forms of expression of the reflexive operational traditions. Initiatives in areas inside and outside the realm of official design will be discussed, as well as dialogic forms of visual mediation within design aimed at the forming of independent opinion and participation.

design

beyond

Design

kritische reflectie & de praktijk van de visuele communicatie

critical reflection & the practice of visual communication

friday 7 & saturday 8 november 1997 symposium

Jan van Eyck Akademie, Maastricht The Netherlands *Speakers* **Susan Buck-Morss,** professor of political philosophy and social theory Cornell University, Ithaca (USA) **Gui Bonsiepe, Cees Hamelink, Heinz Paetzold,** professor of interface design, Cologne professor of international communication professor of philosophy, University of Hamburg University of the Americas, Puebla (Mexico) University of Amsterdam theory department Jan van Eyck Akademie **Sheila Levrant de Bretteville, Jörg Petruschat,** designer, director of graphic design Yale University, New Haven philosopher, design critic «Form+Zweck» Humboldt University Berlin **Michael Rock & Susan Sellers, Jan van Toorn** designer designer, design critic 2x4, New York 2x4, New York designer, Jan van Eyck Akademie

Respondents Dawn Barrett, Andrew Blauvelt, Max Bruinsma, Alex Jordan, Carel Kuitenbrouwer, Tomás Maldonado, Gérard Paris-Clavel, Rick Poynor, Lorraine Wild. Detailed information & reservations from september 1 (including special travel and accommodation arrangements in cooperation with the Dutch Design Institute, Amsterdam). Jan van Eyck Akademie, Academieplein 1 6211KM Maastricht, The Netherlands e-mail vaneyck@xs4all.nl telephone +31.(0)43.350 37 37 fax +31.(0)43.350 37 99

The public programme of the Jan van Eyck Akademie is made possible by the financial support of the Ministry of education, culture and science and the Production fund of the academy with contributions from the Province of Limburg, the Municipality of Maastricht and Drukkerij Rosbeek bv, Nuth

FOREST LODGE PUBLIC SCHOOL,
125
F.L.P.S
1883–2008
ONE HUNDRED
&
TWENTY FIVE YEARS

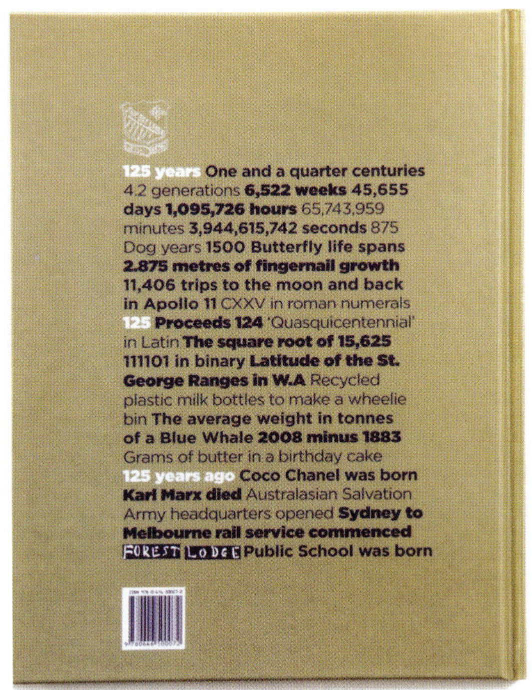

125 years One and a quarter centuries 4.2 generations 6,522 weeks 45,655 days 1,095,726 hours 65,743,959 minutes 3,944,615,742 seconds 875 Dog years 1500 Butterfly life spans 2.875 metres of fingernail growth 11,406 trips to the moon and back in Apollo 11 CXXV in roman numerals 125 Proceeds 124 'Quasquicentennial' in Latin The square root of 15,625 111101 in binary Latitude of the St. George Ranges in W.A Recycled plastic milk bottles to make a wheelie bin The average weight in tonnes of a Blue Whale 2008 minus 1883 Grams of butter in a birthday cake 125 years ago Coco Chanel was born Karl Marx died Australasian Salvation Army headquarters opened Sydney to Melbourne rail service commenced FOREST LODGE Public School was born

Captains' Message
Part One
THE LIFE & TIMES OF OUR SCHOOL

2007 ART SHOW
Part Three
And now we're 125 YEARS YOUNG

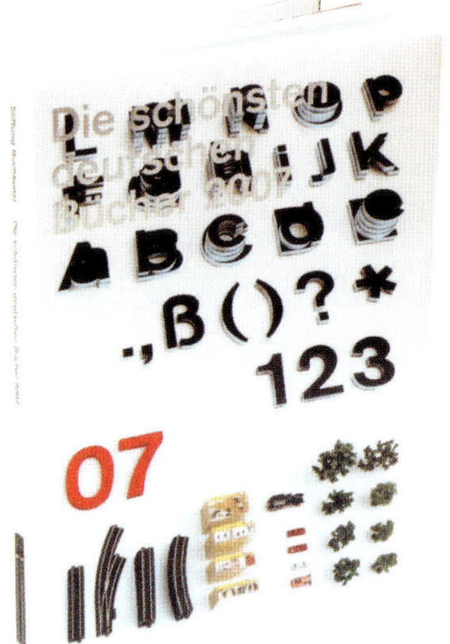

Die schönsten deutschen Bücher 2007
A B C D E
., B ( ) ? *
123
07

BREHMS VERLORENES TIERLEBEN
11

Graphic identity for the Norwegian creative agency Dist.
Focusing on the tools of the creatives such as rough
sketches from the drawing board and the starting point
of a good idea, we combined this creative outcome
with a strict and edgy logo signature and a grid
template of a strange cult who follow a certain set of

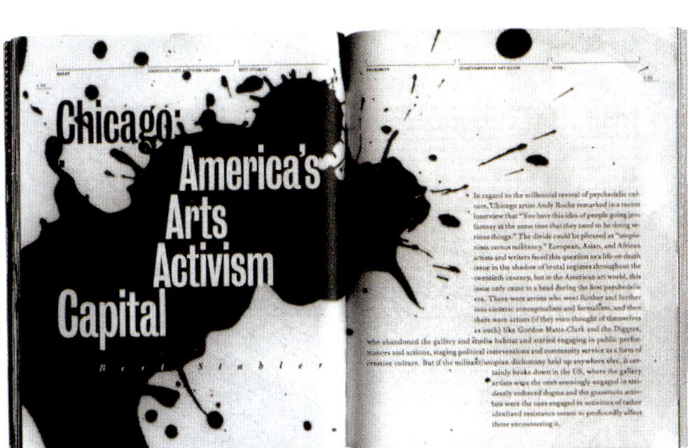

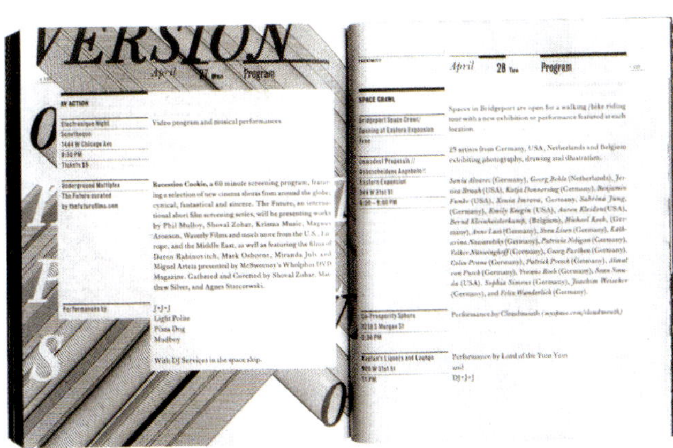

3

5

3.   Exhibition window decoration, *Plakat-
      sammlung*, Museum für Gestaltung
      Zürich, Switzerland, 2005
----
4.   Book cover, *And There Will Be Light*,
      Switzerland, 2006. Self-promotional
      book of unpublished images
----
5.   Identity system, Neue Kunst Halle St.
      Gallen, Switzerland, 2005-7. Includes
      invitations, diaries, exhibition
      catalog, and book

6-7. Exhibition posters, *Animaux, von
      menschen und tieren*, Seedamm
      Kulturzentrum, Schöne Aussichten am
      Bellevue, Switzerland, 2004
----
8.   Book illustration, *Armpit of the
      Mole*, Fundacis 30km/S, France and
      USA, 2005

4

The designer created all visual elements, from posters to
exhibition graphics, for a show about visual cognition
in painting and surgery at the Dulwich Picture Gallery in
London. The specially developed dot-matrix font uses the
dots at different brightness levels to create a sense of
the characters drifting towards and away from the surface.

artwork title
Electronic Flow-Market

typeface
bf_SubZero

designer
Andrea Markewitz

HELLO, THIS IS PONG.

ABOUT PONG IS A BOLD AND CONDENSED TYPEFACE DESIGNED FOR THE USE OF KENNY J. HUANG SELF-IDENTITY. THE GOAL WAS TO PUSH THE LIMIT OF USING MINIMAL PIXELS TO CREATE A LEGIBLE HEADLINE TYPEFACE.

SIZE

20 / 20 — THE NORTHERN CITY IS ALSO ONE OF GERMANY'S MOST IMPORTANT INDUSTRIAL LOCATIONS AND GERNERAL FREIGHT HUBS. SO SWISS'S NEW BASEL-HAMBURG ROUTE WILL LINK TWO OF EUROPE'S PRIME LOGISTICS AND FORWARDING CENTRES WITH DAILY NON-STOP SERVICE.

25 / 25 — THE NORTHERN CITY IS ALSO ONE OF GERMANY'S MOST IMPORTANT INDUSTRIAL LOCATIONS AND GERNERAL FREIGHT HUBS. SO SWISS'S NEW BASEL-HAMBURG ROUTE WILL LINK TWO OF EUROPE'S PRIME LOGISTICS AND FORWARDING CENTRES WITH DAILY NON-STOP SERVICE.

30 / 30 — THE NORTHERN CITY IS ALSO ONE OF GERMANY'S MOST IMPORTANT INDUSTRIAL LOCATIONS AND GERNERAL FREIGHT HUBS. SO SWISS'S NEW BASEL-HAMBURG ROUTE WILL LINK TWO OF EUROPE'S PRIME LOGISTICS AND FORWARDING CENTRES WITH DAILY NON-STOP SERVICE.

40 / 40 — THE NORTHERN CITY IS ALSO ONE OF GERMANY'S MOST IMPORTANT INDUSTRIAL LOCATIONS AND GERNERAL FREIGHT HUBS. SO SWISS'S NEW BASEL-HAMBURG ROUTE WILL LINK TWO OF EUROPE'S PRIME LOGISTICS AND FORWARDING CENTRES WITH DAILY NON-STOP SERVICE.

50 / 50 — THE NORTHERN CITY IS ALSO ONE OF GERMANY'S MOST IMPORTANT INDUSTRIAL LOCATIONS AND GERNERAL FREIGHT HUBS. SO SWISS'S NEW BASEL-HAMBURG ROUTE WILL LINK TWO OF EUROPE'S PRIME LOGISTICS AND FORWARDING CENTRES WITH DAILY NON-STOP SERVICE.

BASIC CHARACTER SET / 100pt

ABCDEFGHI
JKLMNOPQR
STUVWXYZ
abcdefghij
klmnopqrs
tuvwxyz
1234567890
!?{}[]()<>
+-×/=.,;:''""

FORM HAMBERGEFONT hambergefont

95pt u / lc GRRRRRRRRRRRR! Grrrrrrrrrrrrr!

METAMORPH
9. MOSTRA INTERNAZIONALE DI ARCHITETTURA
9. INTERNATIONAL ARCHITECTURE EXHIBITION
12.09.2004–07.11.2004
TUTTI I GIORNI/EVERYDAY 10.00 > 18.00
VENEZIA–GIARDINI/ARSENALE
ENTRATA/ENTRANCE
la Biennale di Venezia

2

METAMORPH
9. INTERNATIONAL ARCHITECTURE EXHIBITION
FOCUS
METAMORPH
9. INTERNATIONAL ARCHITECTURE EXHIBITION
TRAJECTORIES
METAMORPH
9. INTERNATIONAL ARCHITECTURE EXHIBITION
VECTORS
Marsilio

3

PRINCET
ERSITY S
F ARCHI
GRADUA
AM IN A
TURE 20
PRINCETON UNIVERSITY SCHOOL OF ARCHITECTURE.
PROGRAMS OFFICE, 609-258-3641/ FAX: 609-258-474
E-MAIL: SOA@PRINCETON.EDU / DESIGN: OMNIVORE / PR

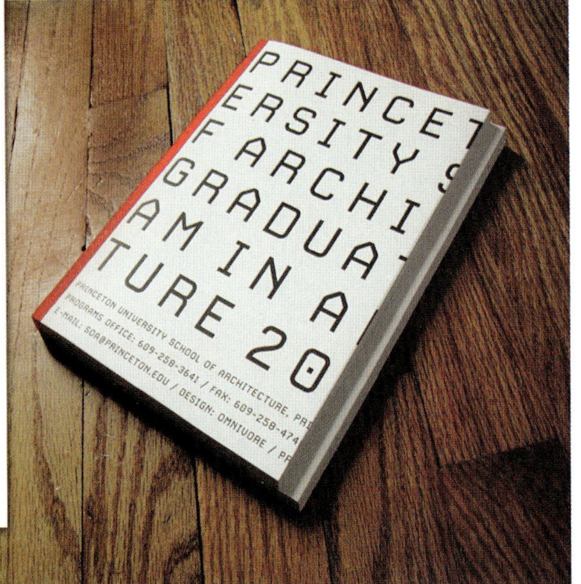

L'Écriture
Il Corsivo
Script
UN SEDICESIMO · numero 6, settembre - ottobre 2008 · Registrazione al tribunale di Mantova n. 4 del 03/02/2007 iscrizione al ROC n. 15649 del 01/08/2007 · Direttore responsabile Federico Maggioni · Progetto a cura di Pietro Corraini · Stampato in Italia a Grafiche SiZ, Verona © 2008 Steven Heller · Maurizio Corraini s.r.l. · via Ippolito Nievo 7a · 46100 Mantova • www.unsedicesimo.com • www.corraini.com • www.aiz.it
ISSN 1972-2842
€ 5.00
Design by Jessica Hische at Louise Fili Ltd.
Un Sedicesimo
SIX
Steven Heller & Louise Fili
Edizioni Corraini

Le Bijou
Acqua
Colonia
PIETRO BORTOLOTTI
BOLOGNA
Senna
Sarius

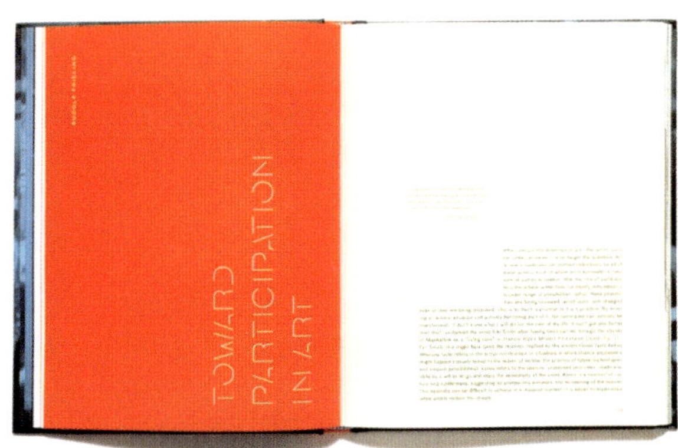

TOWARD
PARTICIPATION
IN ART

A new headline font, Mattias, was created for the launch of this magazine. The all-caps font is very heavy and totally filled in, creating an impression of extreme solidity on the printed page. Using the solid nature of the font, the designers have playfully infilled the letterforms with images to build up a picture (bottom).

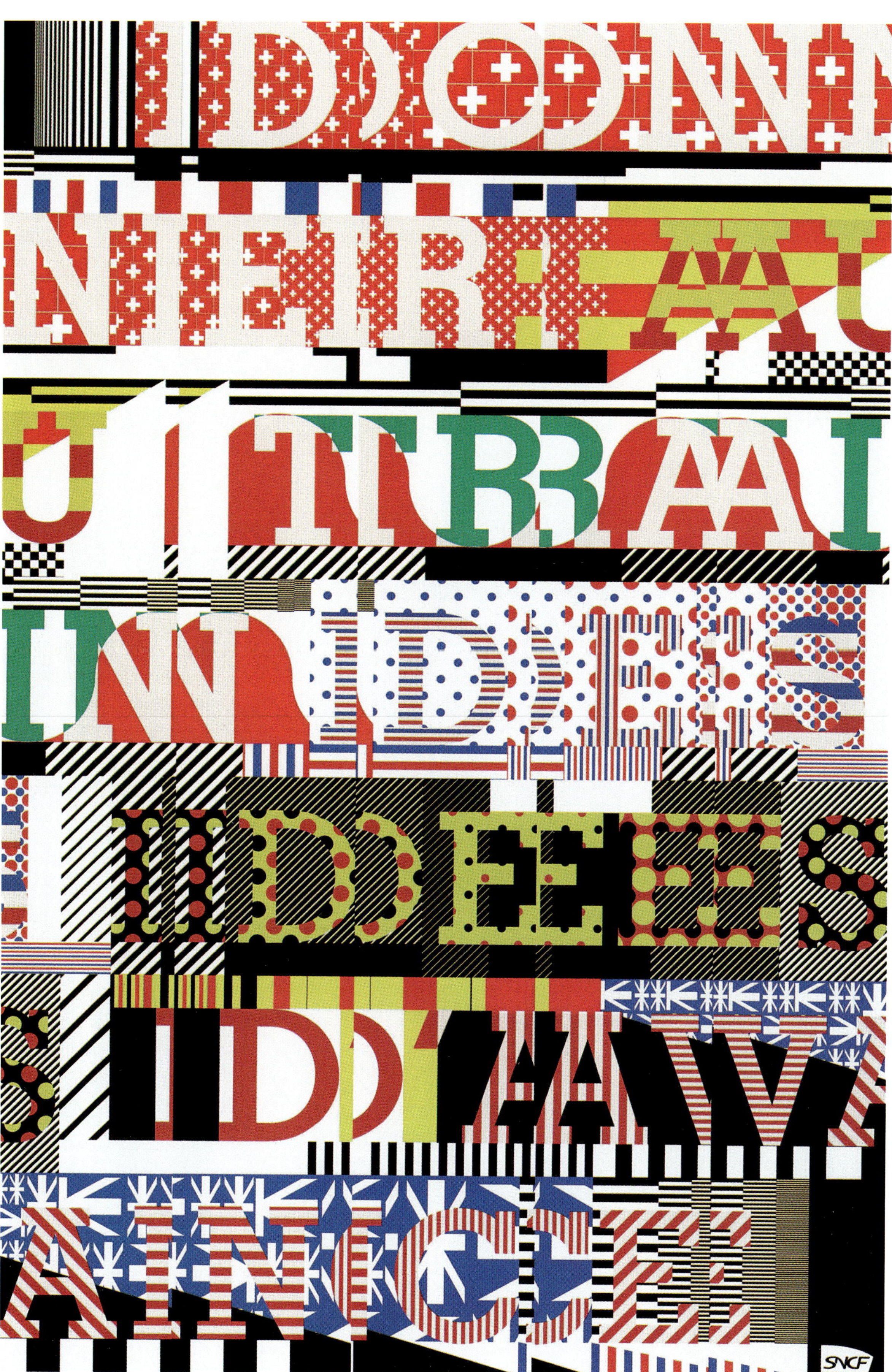

SNCF

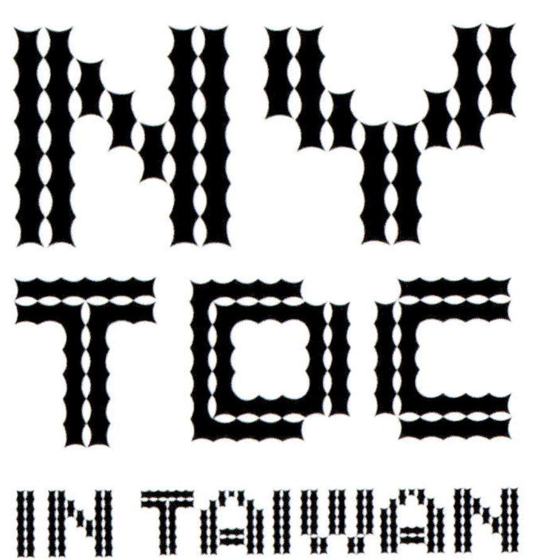

ABCDEFGHIJ
KLNMOPQRST
UVWXYZ !?
1234567890
abcdefghijklmno
pqrstuvwxyz

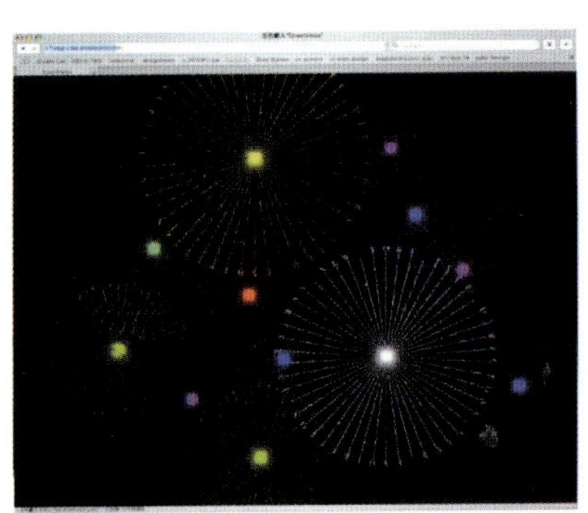

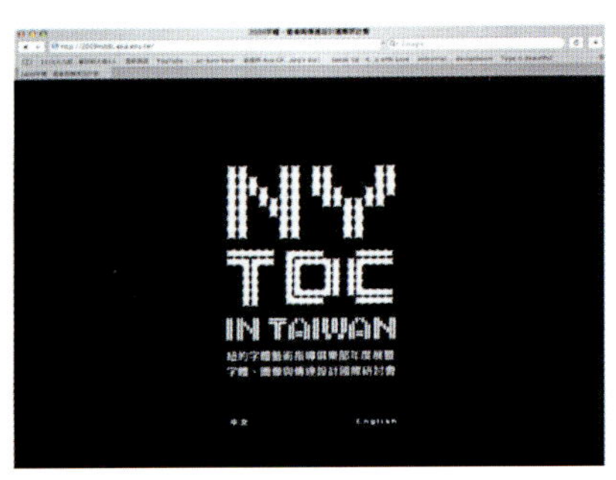

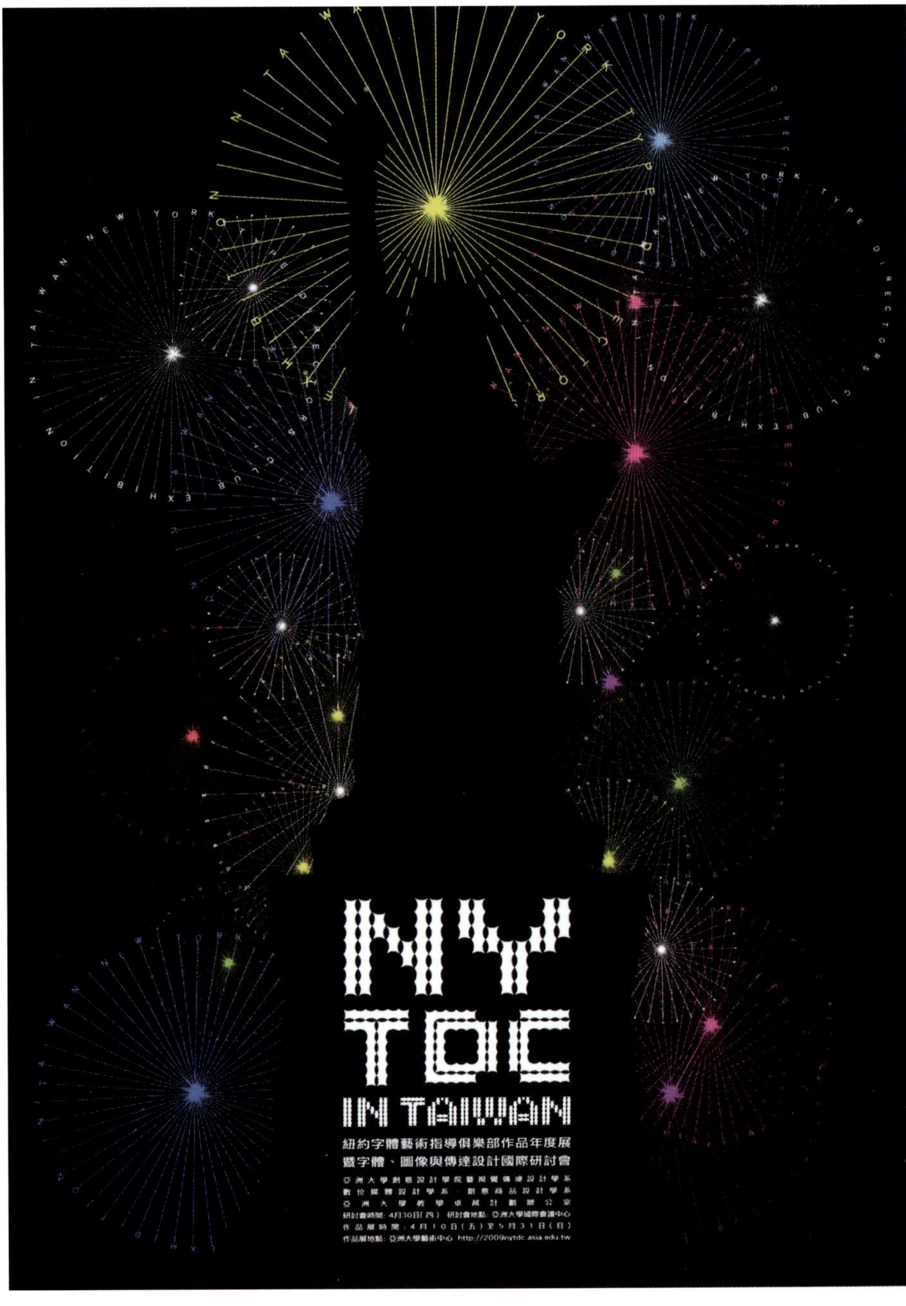

inclination of the slash has been modified in
order to adapt it to the letters of the logotype,
also redesigned to form a compact
unity and diagonal parallel lines.

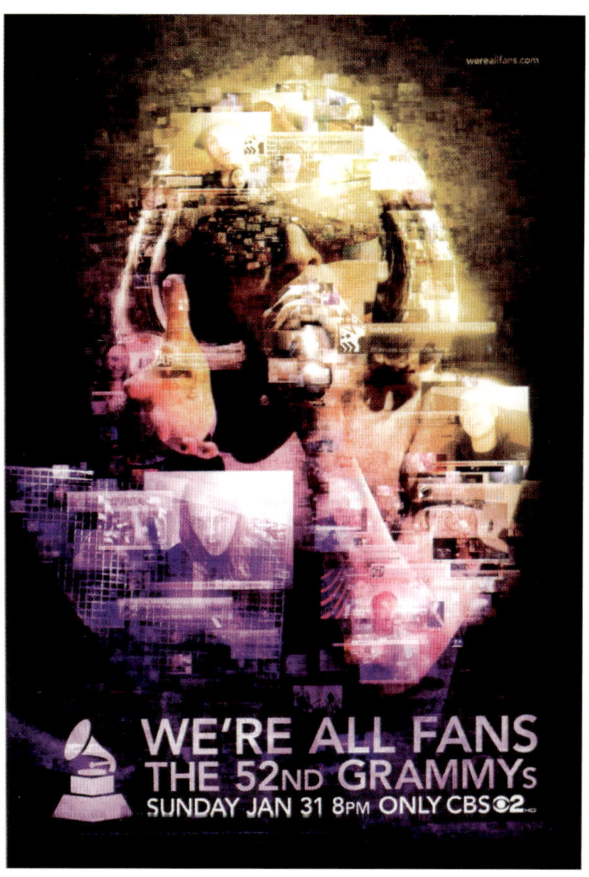

wereallfans.com
WE'RE ALL FANS
THE 52ND GRAMMYS
SUNDAY JAN 31 8PM ONLY CBS 2 HD

wereallfans.com
WE'RE ALL FANS
THE 52ND GRAMMYS
SUNDAY JAN 31 8PM ONLY CBS 2
HDTV

T41 666
krautfromhell
tonite: WE
the last birth of the 1st thing of freaks

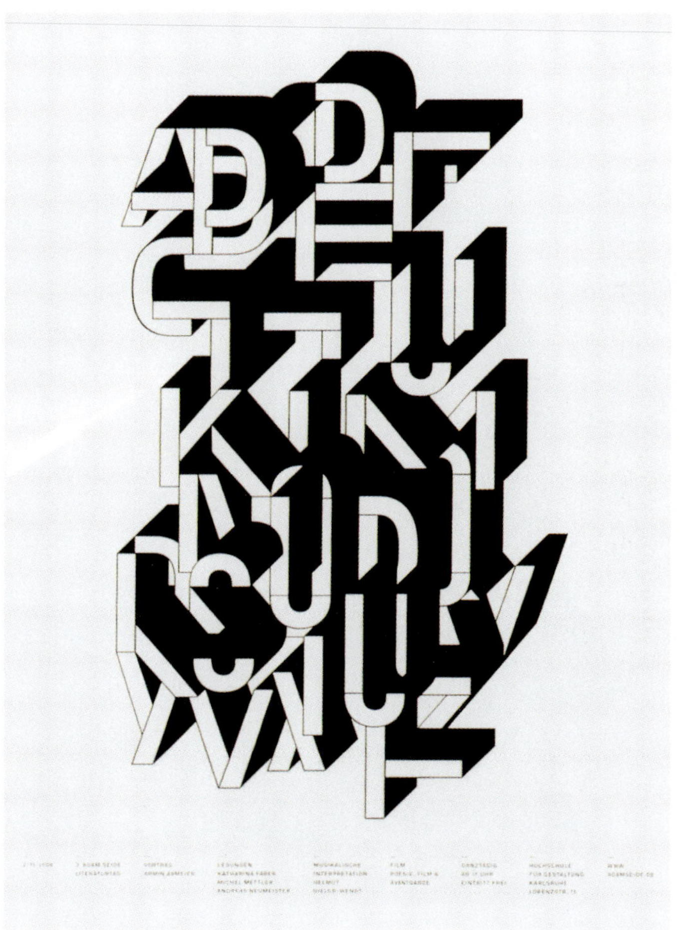

artwork title
TypeFace#4

typeface
BD Fazer

art director
MBrunner

designer
Lopetz

design company
büro destruct

The use of the Splat typeface in
these animation stills illustrates
the gestural and expressive qualities
of everyday speech in a conversation
between two women.

あかさたなはまやらわっ→〜1
いきしちにひみゆり．ゃ（）二2
うくすつぬふむゆる。う「」三3
えけせてねへめ〈〉四4
おこそとのほもよろん．ゃょッォゥャ五六5
アカサタナハマヤラワ6
イキシチニヒミキリヲ7
ウクスツヌフムユル8
エケセテネヘメエレ9
オコソトノホモヨロン10

海浜の碧空に翼舞う

南欧伊太利の紀行文

ミハスの白壁に光が過る

湘南の薔薇園

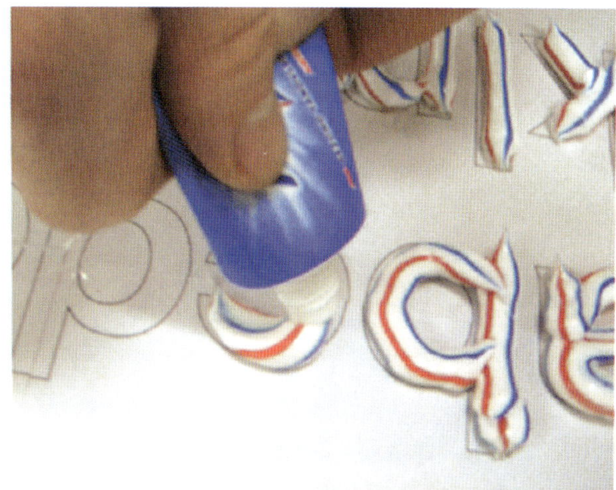

QRS
XYZ

WHERE GREAT PIONEERS MEET

DORNIER
MUSEUM
FRIEDRICHSHAFEN

Le Cirque
MANHATTAN

VERVE
COFFEE ROASTERS

Women SPEAK!

N° 1
Follow the leader
Kick-off nieuwe reeks WomenSpeak!
29.01.08, 8:00-9:30 pm
TUMULT, Domplein 5, Utrecht
Nancy Jouwe, Mira Kho,
Mariëlle Pals, Soula Notos &
Doreen Boonekamp a.v.b.

WOMEN SPEAK!

N° 2
Mama, waar kom ik vandaan?
08.03.08, 2:00-4:30 pm
De Zindering, Stadsschouwburg,
Lucasbolwerk 24, Utrecht
Erika Blikman, René Hoksbergen,
Namrata de Leeuw, Nies Medema,
DJ Nadia, Iris van der Tuin,
Liliane Waanders & Zanillya

Women SPEAK!

N° 6
International sisterhood:
over producenten en consumenten
21.10.08, 8:00-9:30 pm
TUMULT, Domplein 5, Utrecht
Geert-Jan Davelaar, Astrid Kaag,
DJ Nadia, Soula Notos,
Maaike Schouten & Zanilliya

WOMEN SPEAK!

N° 3
Who cares?
20.04.08, 3:00 pm, Toneelstuk
Meisjesstad van STUT Theater
& 4:30-6:00 pm, debat
Stefanus, Amazonedreef 44, Utrecht
Vincent Duindam,
Mariette Hermans, Insos Ireeuw,
Emely Nobis & Astrid Runs

WOMEN SPEAK!

N° 4
Trots op homo's en lesbo's?
De Nederlandse leeuw
in een roze hemdje
15.06.08, 8:00-9:30 pm
TUMULT, Domplein 5, Utrecht
Ossama Abu Amar, Sarah Bracke,
Frank van Dalen,
Sooreh Hera & Markha Valenta

WOMEN SPEAK!

N° 5
Migrantenvrouwen:
de nieuwe globalisten
30.09.08, 8:00-9:30 pm
TUMULT, Domplein 4, Utrecht
Izaline Calister,
Stephanie Mbanzendore,
Fatumo Farah, Fenna Ulichki &
Joke van der Zwaard

typeface
Seviche

typeface family
Seviche

designer
Gerry Chapleski

foundry/supplier
Editable Graphics

country of origin
USA

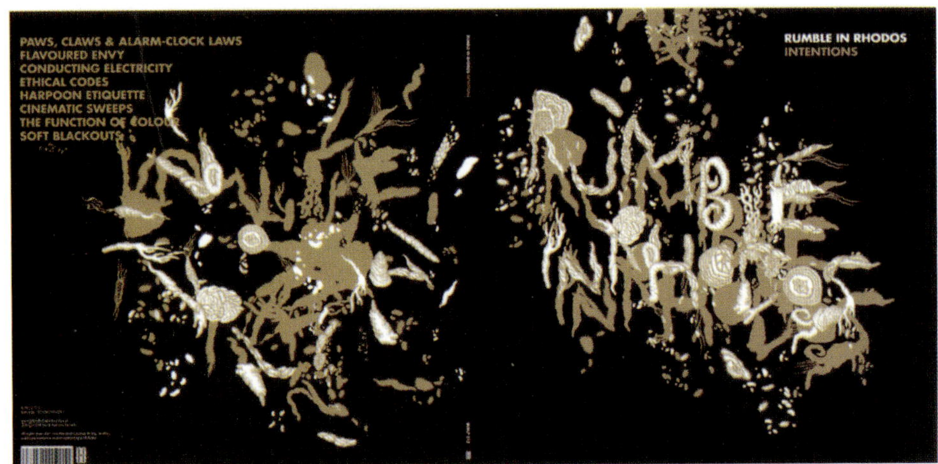

La Reina Maya

Hacienda SAN LUCAS

VOLVO
PENTA
UND THE SA-
MUEL JACKSON
5 AUS NORWE-
GEN SPIELEN
AM 30. OKTO-
BER 2008 UM
20 UHR INSTRU-
MENTALMUSIK
IM STEINBRUCH
IN DUISBURG.

—
L2M3
Ina Bauer, Frank Geiger, Sascha Lobe,
Thorsten Steidle

Schönste deutsche Bücher 2007

Publication presenting a journey through
the German bookscape. Laid out next to
each other, the characteristic style of layout
becomes visible, while details are shown
in large format. Set pieces from the world
of model railways create enactments that
explore the content of the various books.

Client / Publisher Stiftung Buchkunst Frankfurt
am Main und Leipzig
Editor Uta Scheider
Credits Photography by Ina Bauer, Thorsten Steidle,
supported by Wolfram Palmer

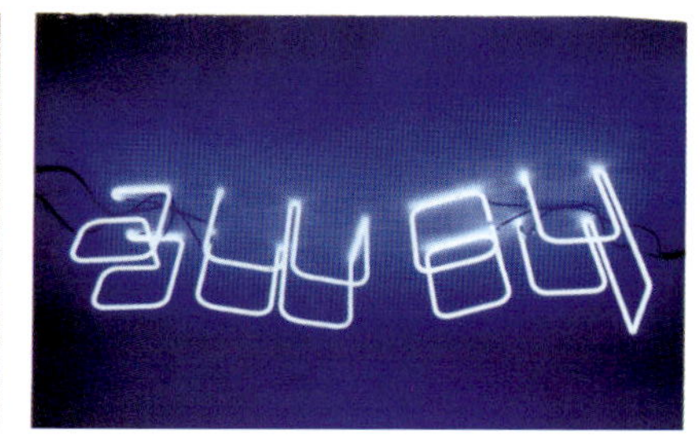

typeface

Splat

typeface family
Splat

designer
Garry Waller

foundry/supplier
Garry Waller

country of origin
UK

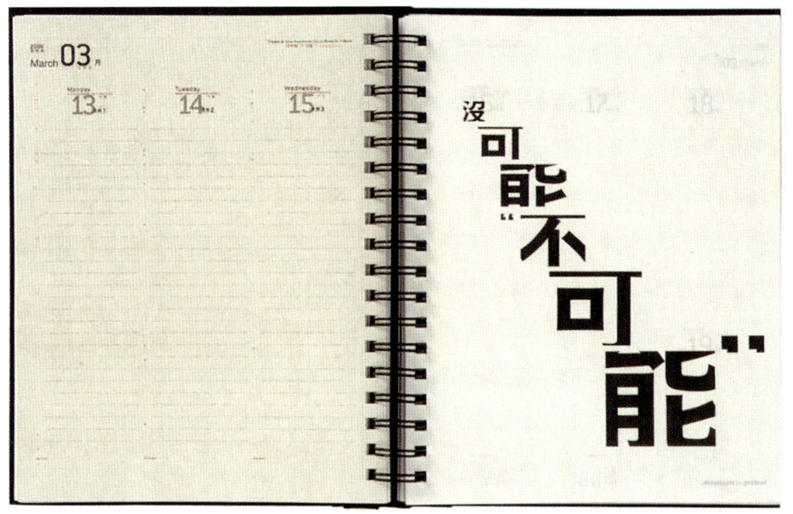

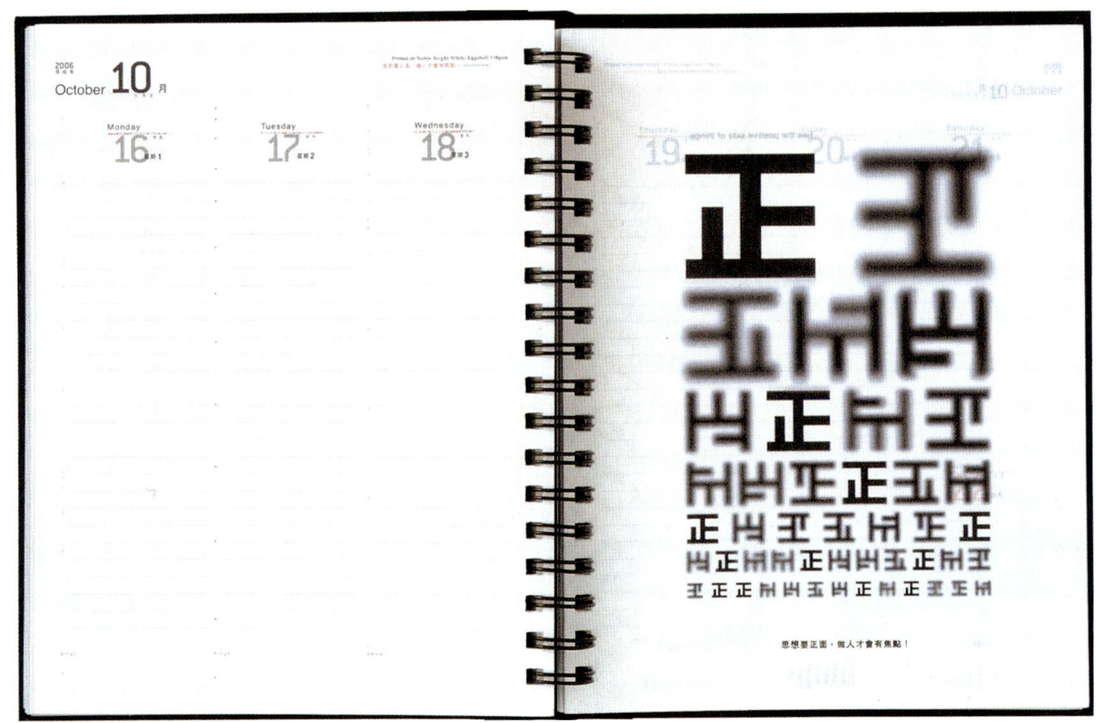

A new headline font, Mattias, was created for the launch
of this magazine. The all-caps font is very heavy and totally
filled in, creating an impression of extreme solidity
on the printed page. Using the solid nature of the font, the
designers have playfully infilled the letterforms with images

WUSEL
BIRGER
ICH
PASSE NICHT SO
IN DIESE
WAHNSINNIGE
FEIGE
ENTSETZLICHE
GESELLSCHAFT
SIE IST SO
VERQUER
DIESE AUSWAHL
EISERNER
ASSE
TOTAL OHNEMICH
EIN FIASKO EIN
CHAOS
ZITTERND ÄHNELT
DIESE WEITLÄUFIGE
EINER TEILWEISE
TOTEN
RUINENSTADT
ICH WIRKE MIT ALS
ZERSTÖRER
UND EIN
WIDERLICHER
UNSINNMAGAGER
DER
EINE EINZIGE
EISERNE EINSAME
EIFERT
EIN ERSTES
EINZIGES EISERNES
WERTVOLLES
GEDANKEN GEDICHT
ZU
ERFINDEN

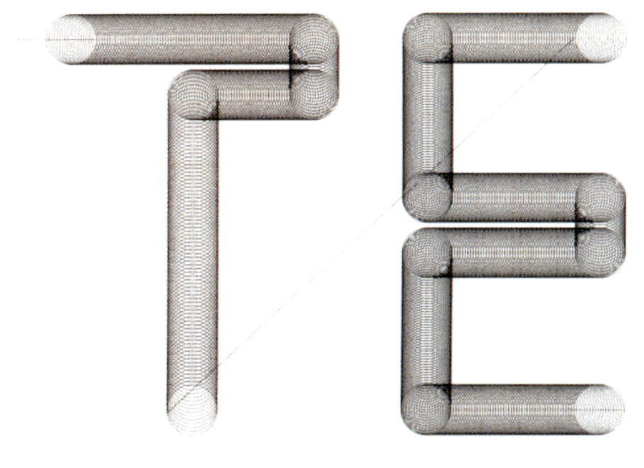

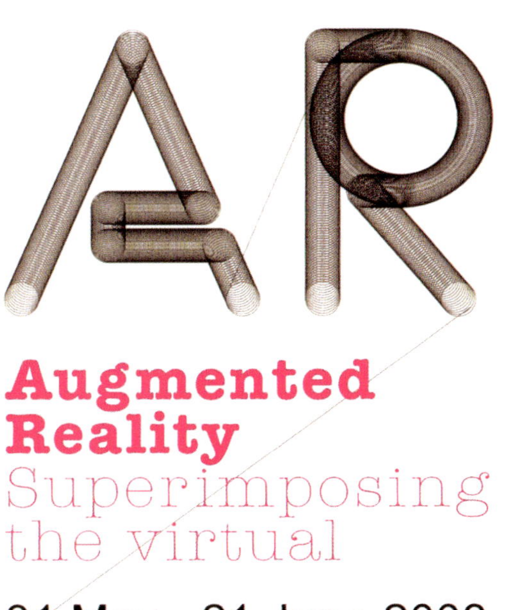

Augmented Reality
Superimposing the virtual

31 May - 21 June 2008

typeface

Futures

typeface family
Futures

designer
Stephen Banham

foundry/supplier
The Letterbox

country of origin
Australia

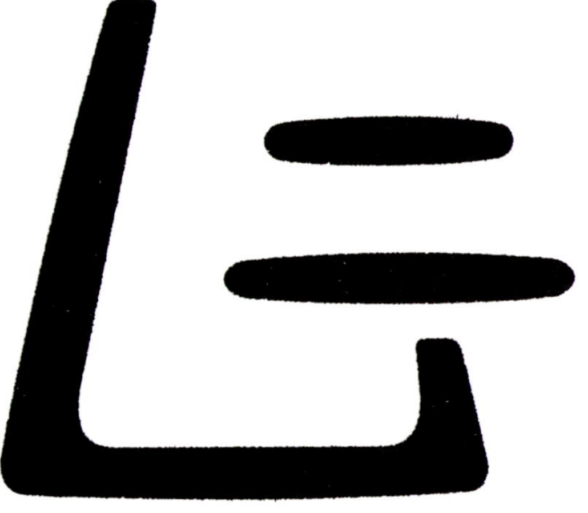

The name and theme of the store is inspired by the pizza trays, which are 80cm by 20cm in size. The concept of the pizza trays are followed through in the signage and menu application.
Visual Unity designed the identity, environmental signage, menus and uniforms.

CHANGE
PROGRESS
HOPE
DREAM
EMPOWER
DREAM
CHANGE
HOPE
UNITE
BELIEVE
HOPE
RESPECT
INCLUDE
DREAM
PROGRESS
HOPE
CHANGE
UNITE
PROGRESS
HOPE
DREAM
CHANGE
UNITE
HOPE
INCLUDE
UNITE
DREAM
RESPECT
INCLUDE
DREAM
HOPE
DREAM
INCLUDE
HOPE
BELIEVE
RESPECT
EMPOWER
BELIEVE
HOPE
HOPE
EMPOWER
PROGRESS
INCLUDE
BELIEVE
DREAM
HOPE
CHANGE
UNITE
BELIEVE
HOPE
HOPE
CHANGE
HOPE
UNITE
UNITE
BELIEVE
HOPE
BELIEVE
HOPE
DREAM
RESPECT
HOPE
INCLUDE
BELIEVE
HOPE PROGRESS
RESPECT
HOPE
INCLUDE
BELIEVE
UNIT
RESPECT
EMPOWER
PROGRESS
EMPOWER
HOPE
RESPECT
EMPOWER
DREAM
PROGRESS EMPOWER
DREAM
BELIEVE
UNITE DREAM
UNITE

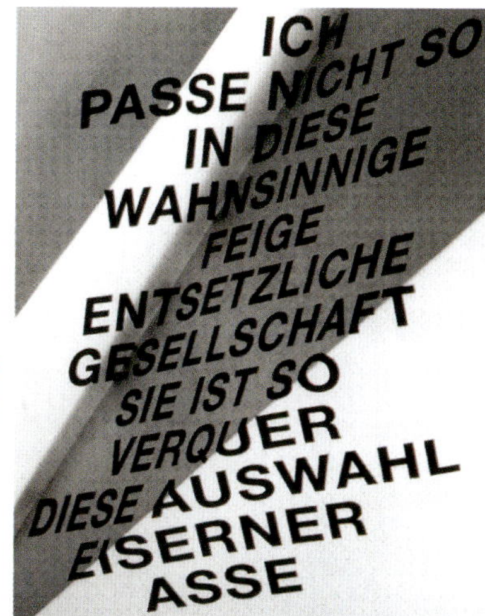

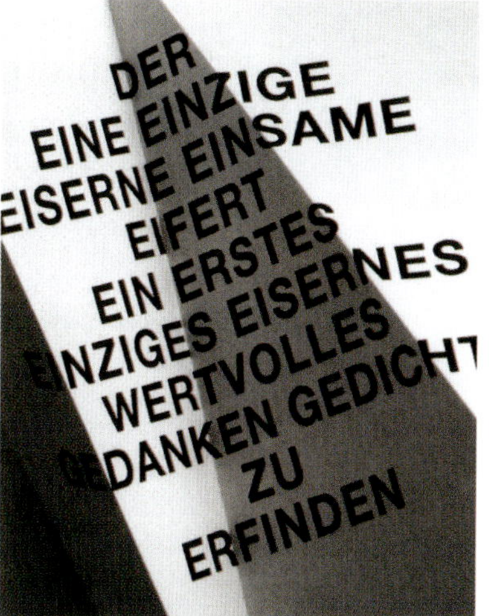

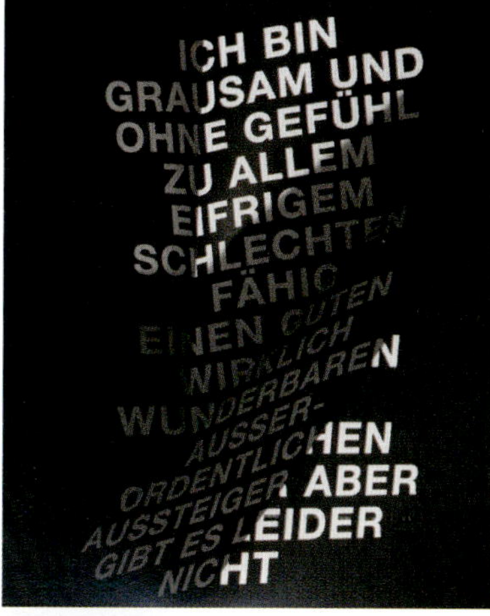

Typeworkshop.com originates from workshops given by Underware, a graphic design studio based in Holland and Finland that specializes in designing and producing typefaces. The workshops help the participants to understand typography more fully in a physical sense. 'Movable Type' used cardboard boxes as its construction material. 'Manual Pixelism' created pixel-based fonts from repeated modules, such as disposable plastic drinking cups, supermarket shopping trolleys and paperback books. 'Let It Run', meanwhile, was constructed as the world's biggest type-domino.

EXPOSITION
NOBEL
au service
de l'innovation
du 7 octobre 2008
au 11 janvier 2009
palais
de la découverte
Avenue Franklin Roosevelt
75008 Paris
T. 01 56 43 20 20
www.palais-decouverte.fr

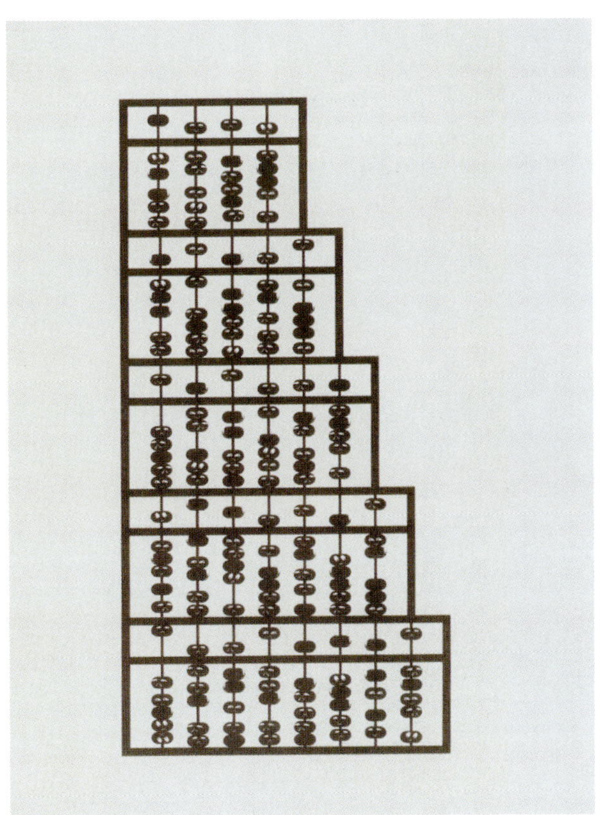

Hoppla... jetzt erst recht!
We can!
äxcüsi
tiptop
pack's!
remember
super
wow!
dänk dra
No Problem
forget it!
weisch? no?
Sh...
krrr...
yeah!
Let's do it!
Neus!
Es guets
...ups
zack!
bis bald!
FUCK

Chic is a Spanish glossy women's monthly. The art director has used a bold curvaceous serif font to create large playful graphic opening spreads for features, sometimes bleeding the letterforms off the page to create semi-abstract shapes and at other times simply setting the title as large as possible while retaining legibility.

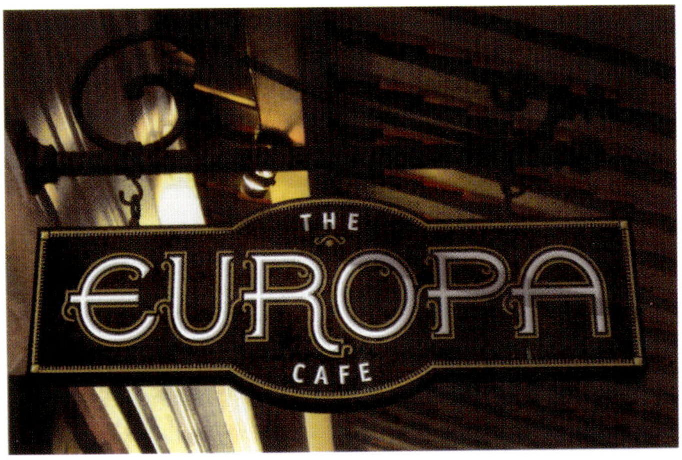

560 MAIN STREET
STROUDSBURG, PA
18360

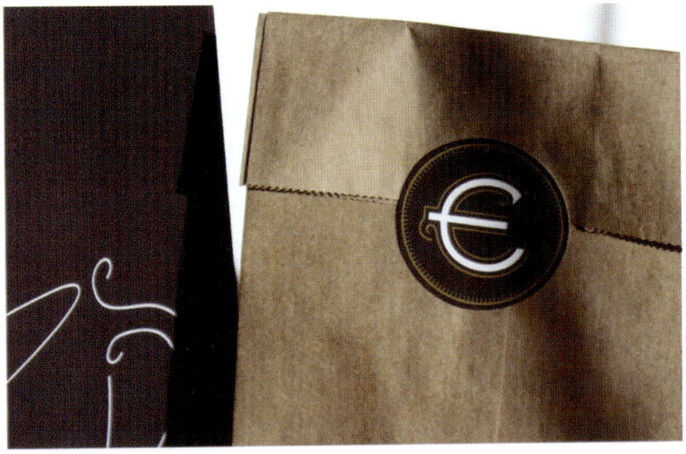

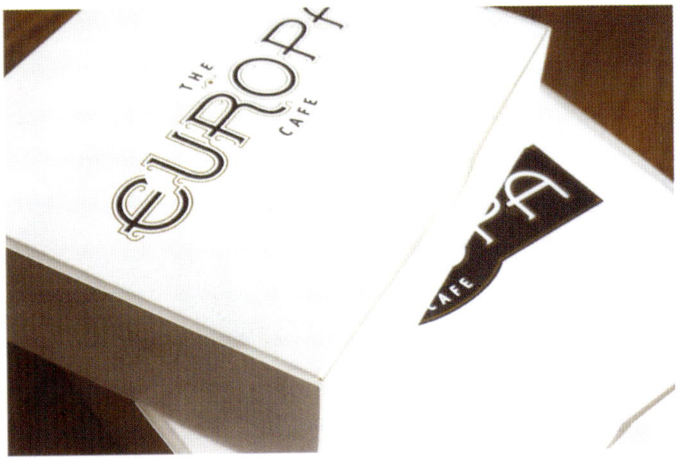

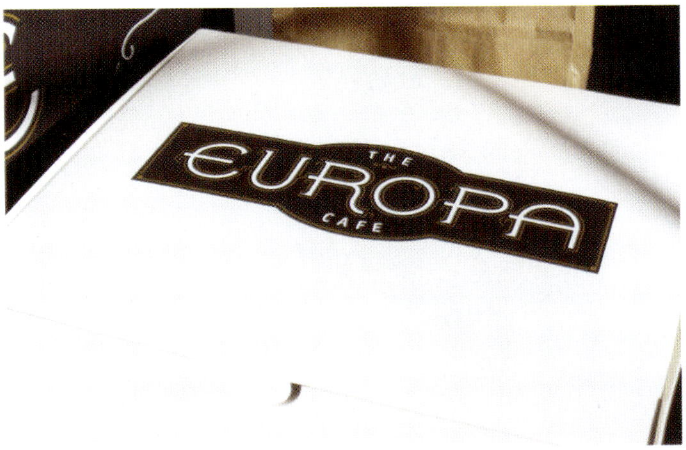

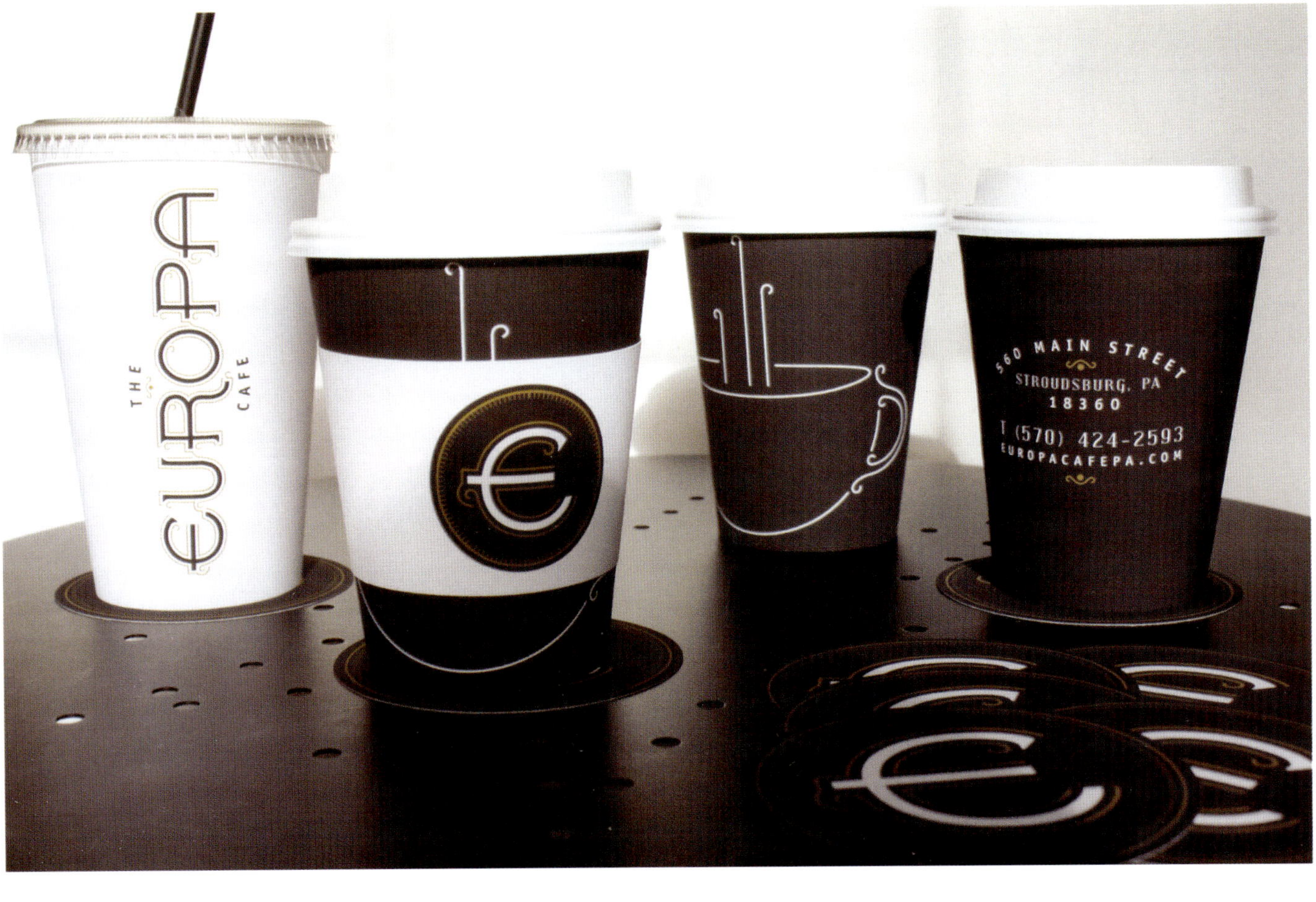

T (570) 424-2593
EUROPACAFEPA.COM

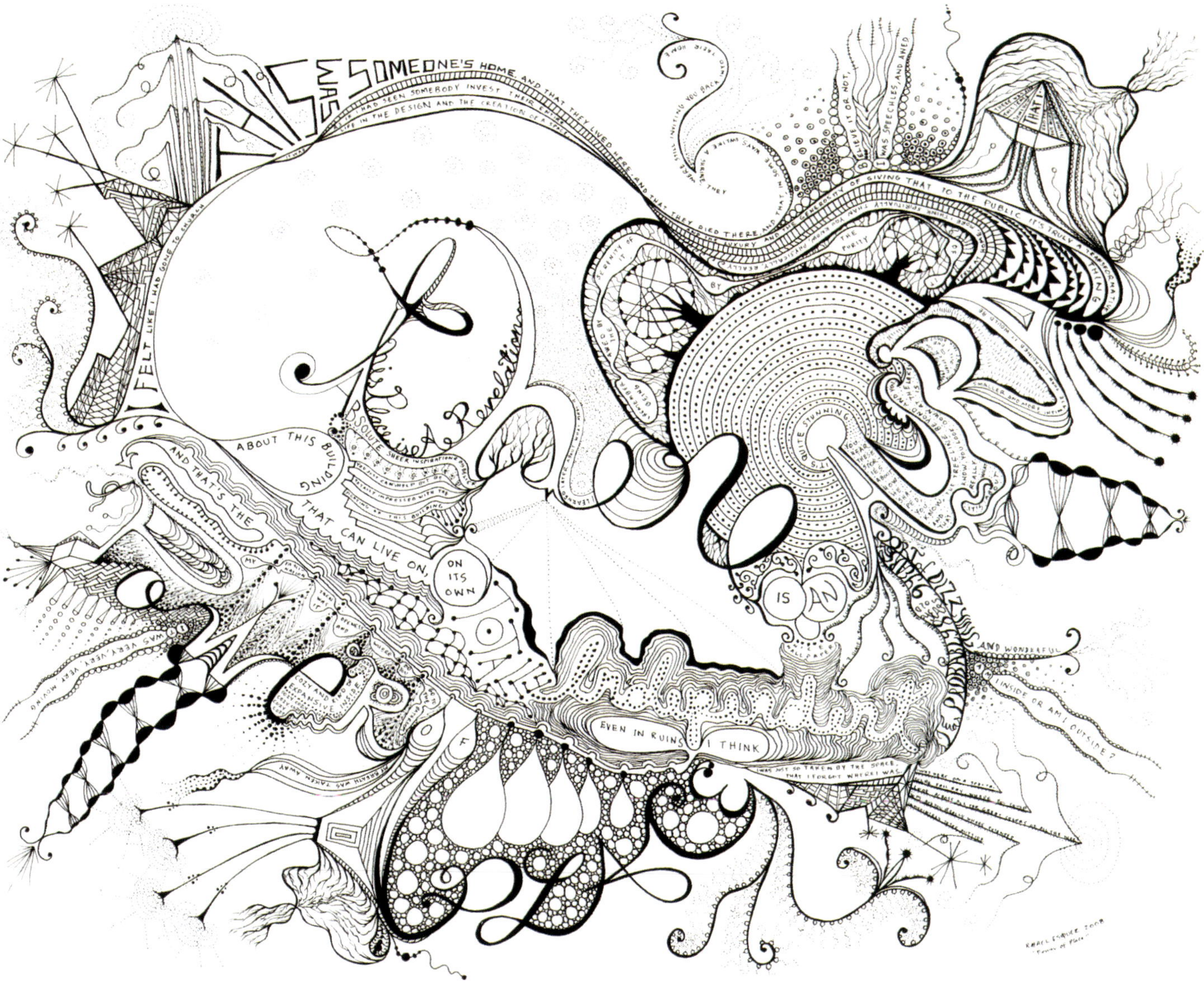

STAY
COOL

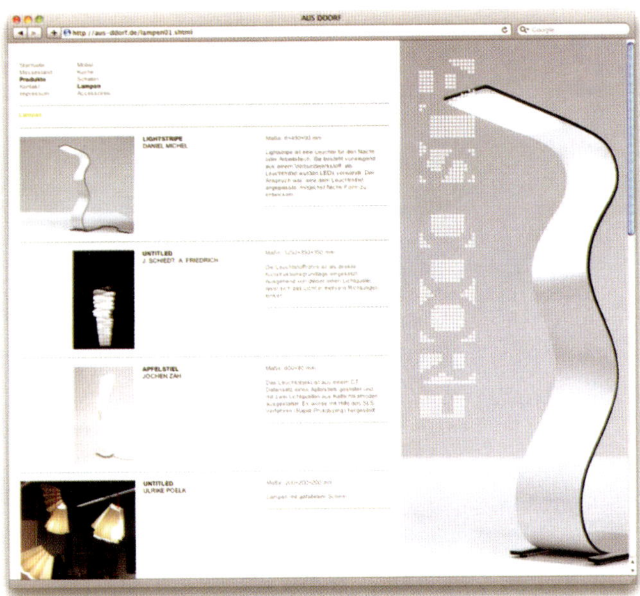

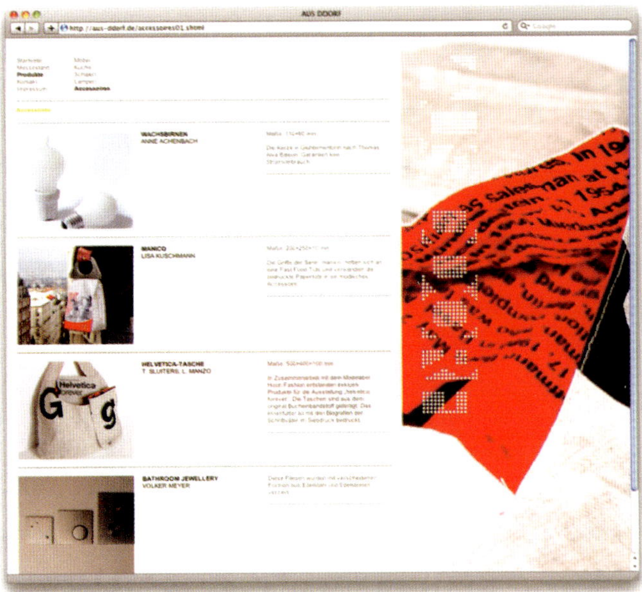

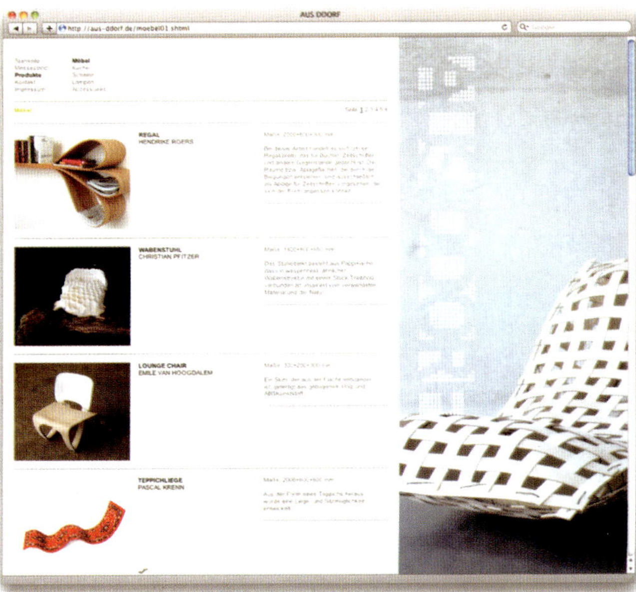

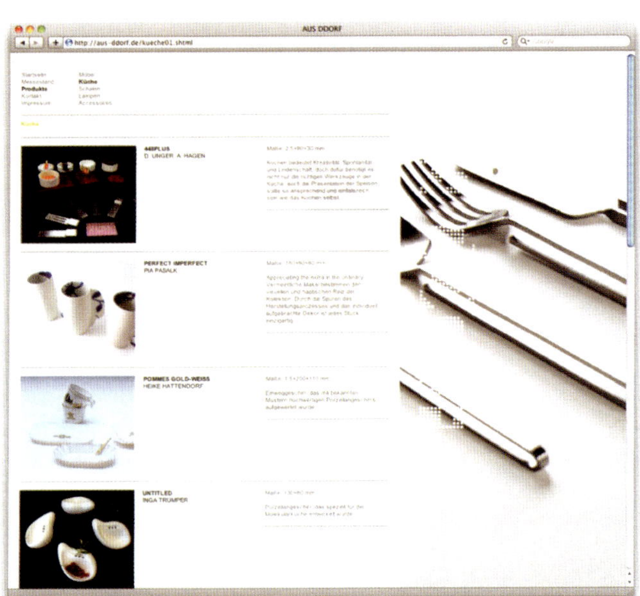

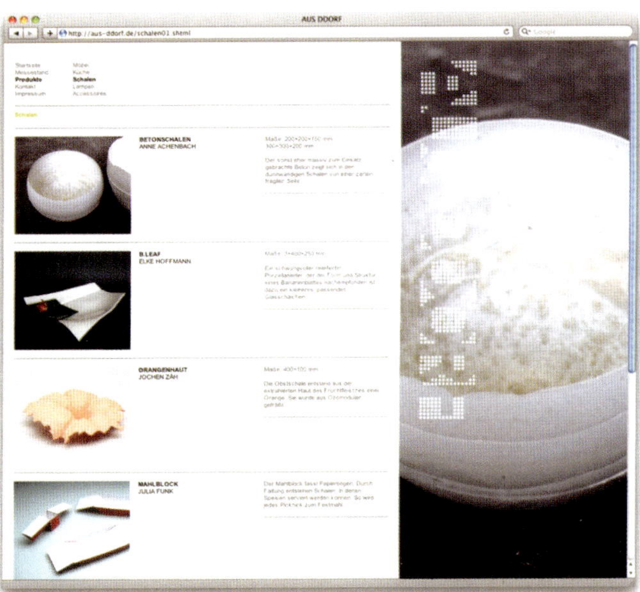

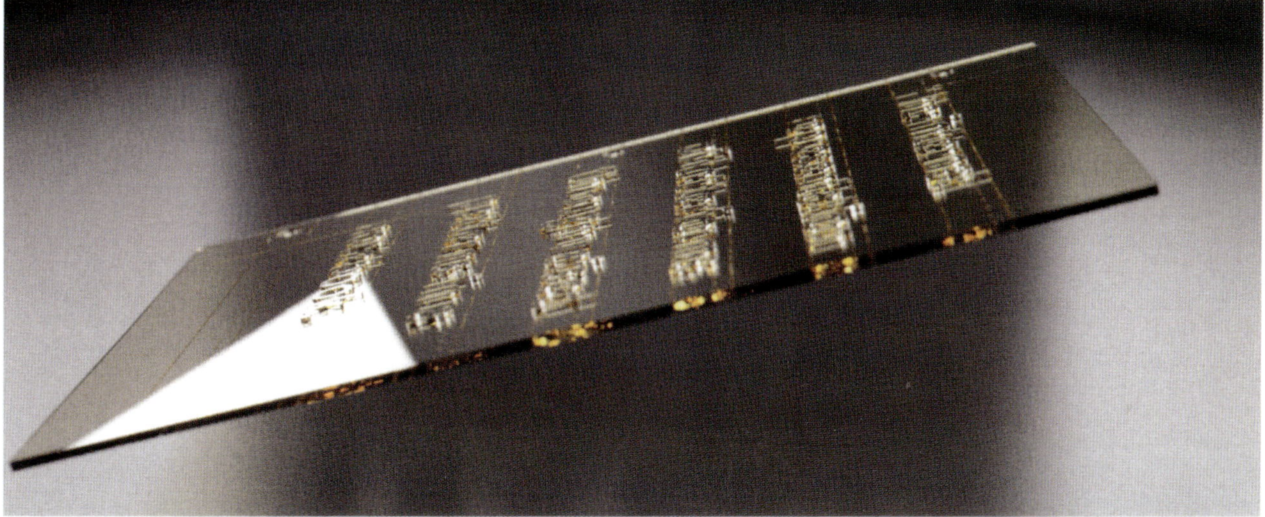

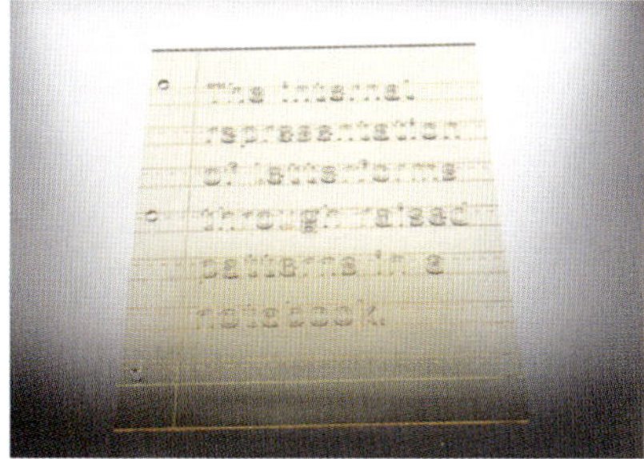

The internal
representation
of letterforms
through raised
patterns in a
notebook.

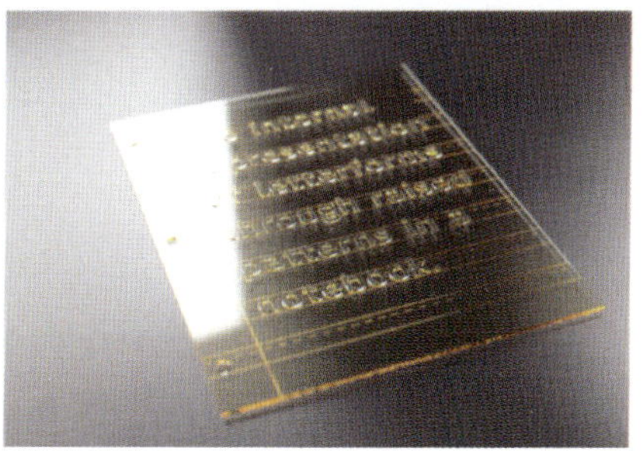

The internal
representation
of letterforms
through raised
patterns in a
notebook.

Stedelijk Museum CS
SMCS
SMCS
SMCS
SMCS

the furniture collection
16.05 - 03.10
20/20 Vision
Meubelindustrie Gelderland BV
Mondriaan Stichting
Stedelijk Museum CS
CS

Stedelijk Museum CS
16.05 31.12
Tussenstand:
een keuze uit de
collectie
Intermission:
a choice from the
collection
16.05 03.10
20/20 Vision
Yesim Akdeniz Graf,
Francis Alÿs, Marc
Bijl, Germaine Kruip,
De Rijke/De Rooij,
Mathias Poledna,
Steve McQueen,
Torbjørn Rødland
16.05 29.08
Kramer vs. Rietveld
/Contrasten in
de meubelcollectie
Kramer vs. Rietveld
/Contrasts in the
furniture collection
www.stedelijk.nl
Stedelijk Museum CS
Oosterdokskade 5
1011 AD Amsterdam
Dagelijks open: 10-18 uur
Donderdag: 10-21 uur
Open daily: 10 am - 6 pm
Thursday: 10 am - 9 pm
Verwacht: Best Verzorgde
Boeken, Prix de Rome,
Sandberg, Gemeente Kunst
Aankopen, Who If Not We,
Robert Smit & Gold, IDFA
Expected: Best Book Design,
Prix de Rome, Sandberg,
Municipal Art Acquisitions,
Who If Not We, Robert Smit
& Gold, IDFA
Meubelindustrie
Gelderland BV
Mondriaan Stichting

Stedelijk Museum CS
SMCS
Stedelijk Museum CS
SMCS

20 20
vision
20 20
vision
20 20
vision
20 20
vision
20 20
vision
Yesim Akdeniz Graf
Francis Alÿs
Marc Bijl
Germaine Kruip
Steve McQueen
Mathias Poledna
De Rijke/De Rooij
Torbjørn Rødland
16.05 - 03.10.2004

20 20
vision
Yesim Akdeniz Graf   Steve McQueen   16.05 - 03.10.2004
Francis Alÿs   Mathias Poledna   Stedelijk Museum CS
Marc Bijl   De Rijke/De Rooij   Oosterdokskade 5
Germaine Kruip   Torbjørn Rødland   1011 AD Amsterdam
20 20
SMCS
vision

typeface

AF Hadrian Roman

typeface family
AF Hadrian

designer
Christian Küsters

foundry/supplier
ACME fonts

country of origin
UK

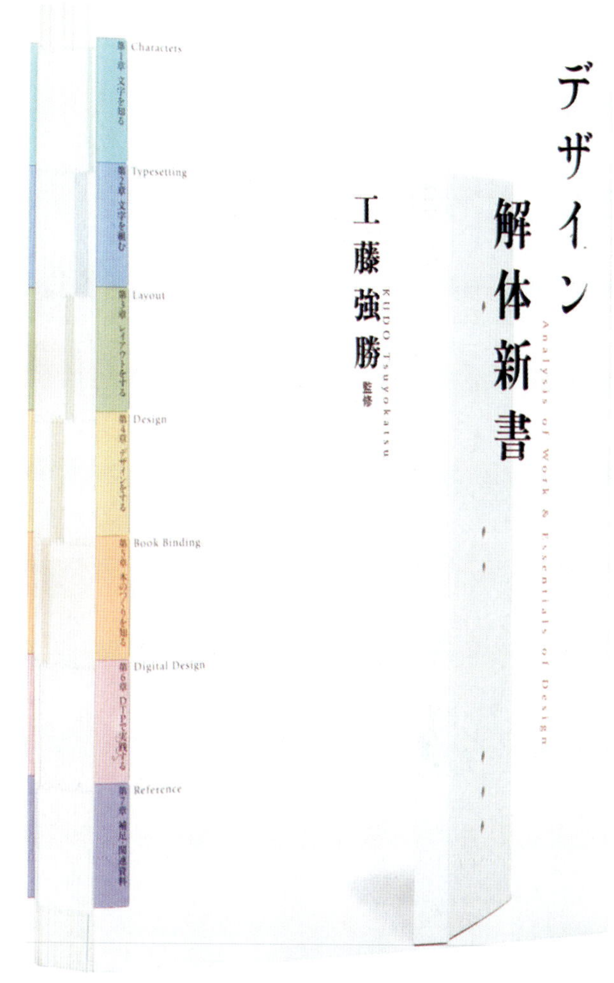

Circle of life

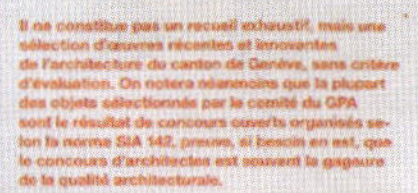

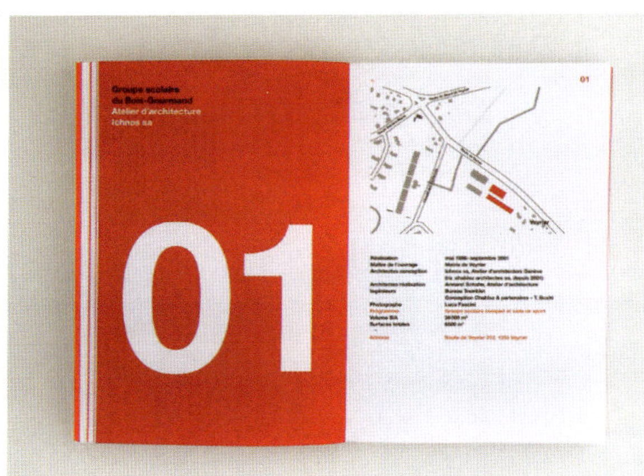

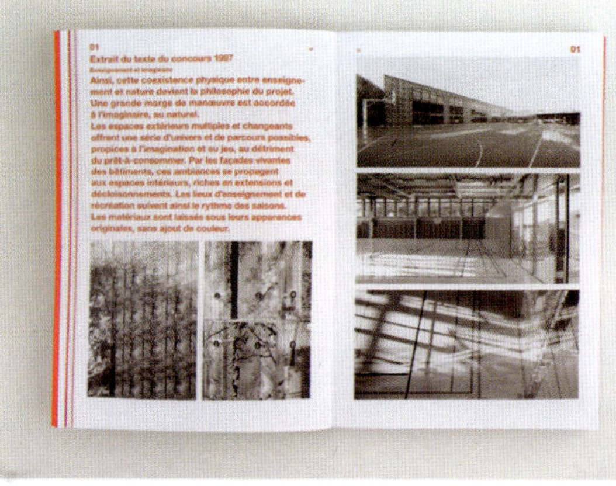

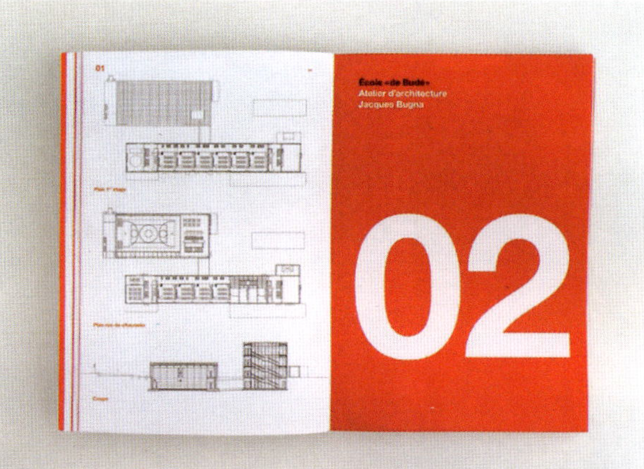

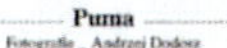

Puma
Fotografie _ Andrzej Dudosz

Karuzela
Fotografie _ Andrzej Dudosz

typeface

AF Hadrian Roman

typeface family
AF Hadrian

designer
Christian Küsters

foundry/supplier
ACME fonts

country of origin
UK

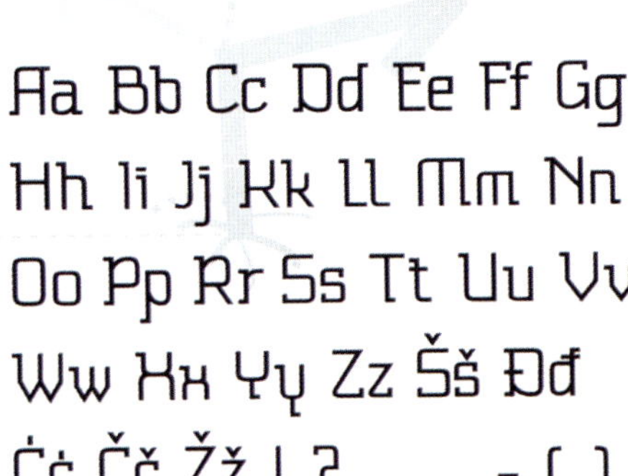

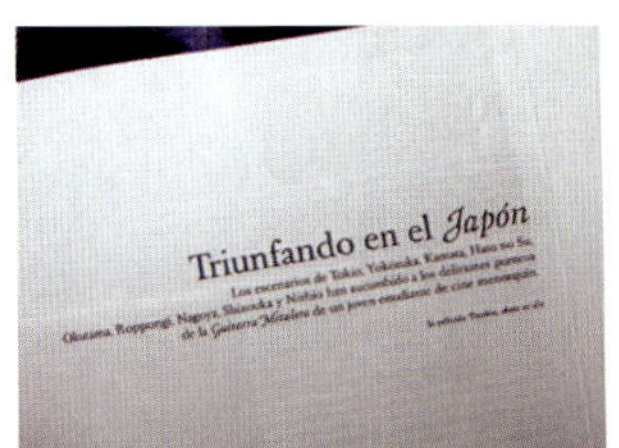

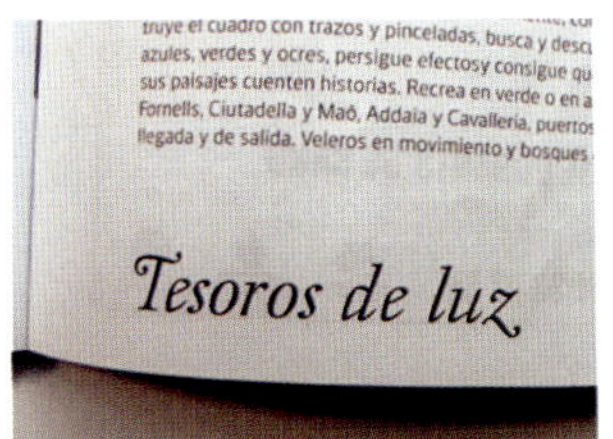

2/11/2008
LITERATURTAG

3. ADAM SEIDE

VORTRAG
ARMIN ABMEIER

LESUNGEN
KATHARINA FABER
MICHEL METTLER
ANDREAS NEUMEISTER

MUSIKALISCHE
INTERPRETATION
HELMUT
BIELER-WENDT

FILM
POESIE, FILM &
AVANTGARDE

GANZTÄGIG
AB 11 UHR
EINTRITT FREI

HOCHSCHULE
FÜR GESTALTUNG
KARLSRUHE
LORENZSTR. 15

WWW.
ADAMSEIDE.DE

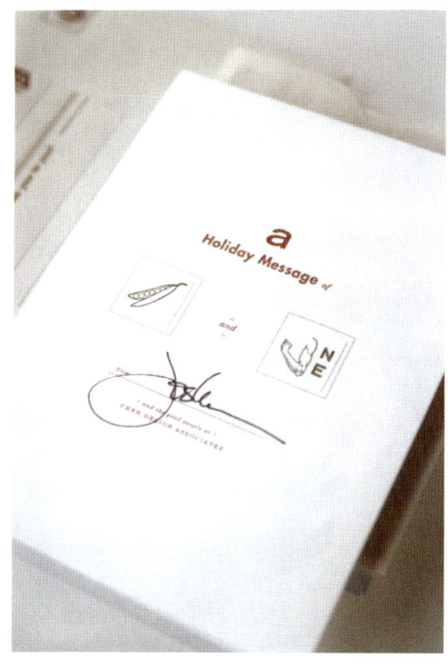

a
Holiday Message of
and

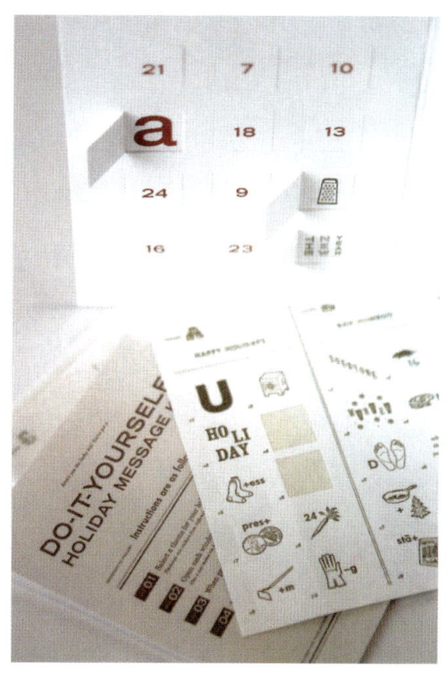

21 7 10
a 18 13
24 9
16 23
HAPPY HOLIDAYS
U
HO LI
DAY
DO-IT-YOURSELF
HOLIDAY MESSAGE

A font collection that explores different typographic
qualities and possibilities when used in print and on screen.
The interactive application was finished in early January
2001 for the Royal College of Art 'Work in Progress'
exhibition. There are three typefaces: Delayed, Unfolded
and Binary. Each is available in different sound and
animation modes. By using mouse and keyboard, you can
play and write with the application. The AAT application
can be downloaded for free at www.flat33.com.

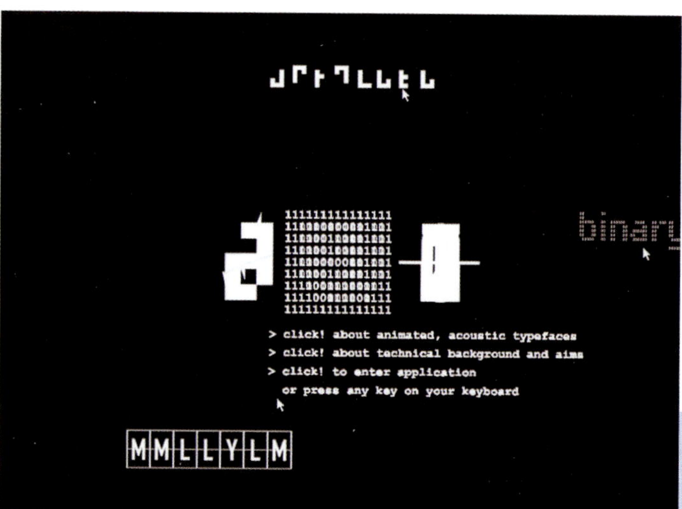

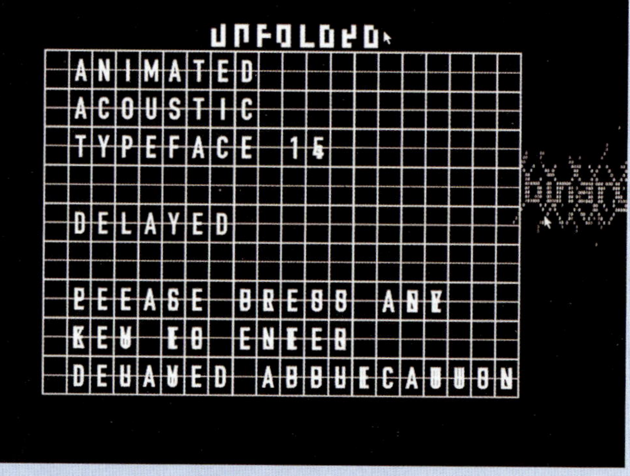

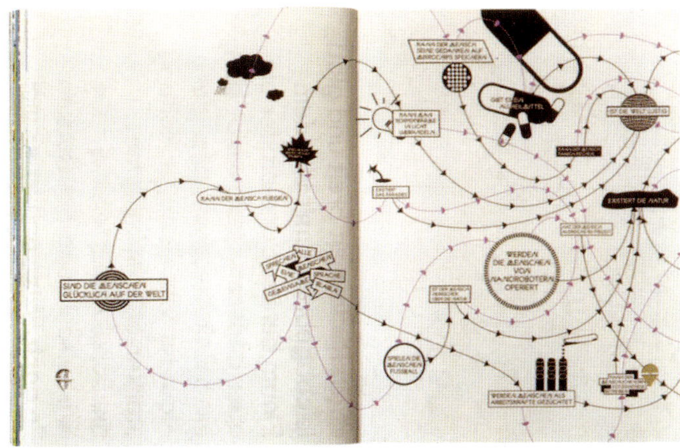

CONCH
REPUBLIC
Conch
23.04.1982
Dennis Wardlow
Prime Minister
Dennis Wardlow
Republik
15.4 km²
24.768
Englisch
Conch Dollar
Amerika
UTC-5
24°33'33.00"N
81°47'02.51"W
167/199+X
PRINCIPALITY
OF SEALAND
012/199+X
Sealand
02.09.1967
UTC+0
Paddy Roy Bates
Prince Regent
Fürstentum
54°41'03.82"N
25°18'23.93"E
Englisch
Sealand Dollar
Europa

UTOPIE
DREHT SICH DIE ERDE

verzogen wird: wie
eines Segelschiffes,
zwischen den Seilen,
ihn halten, nicht richt
aufgespannt ist.

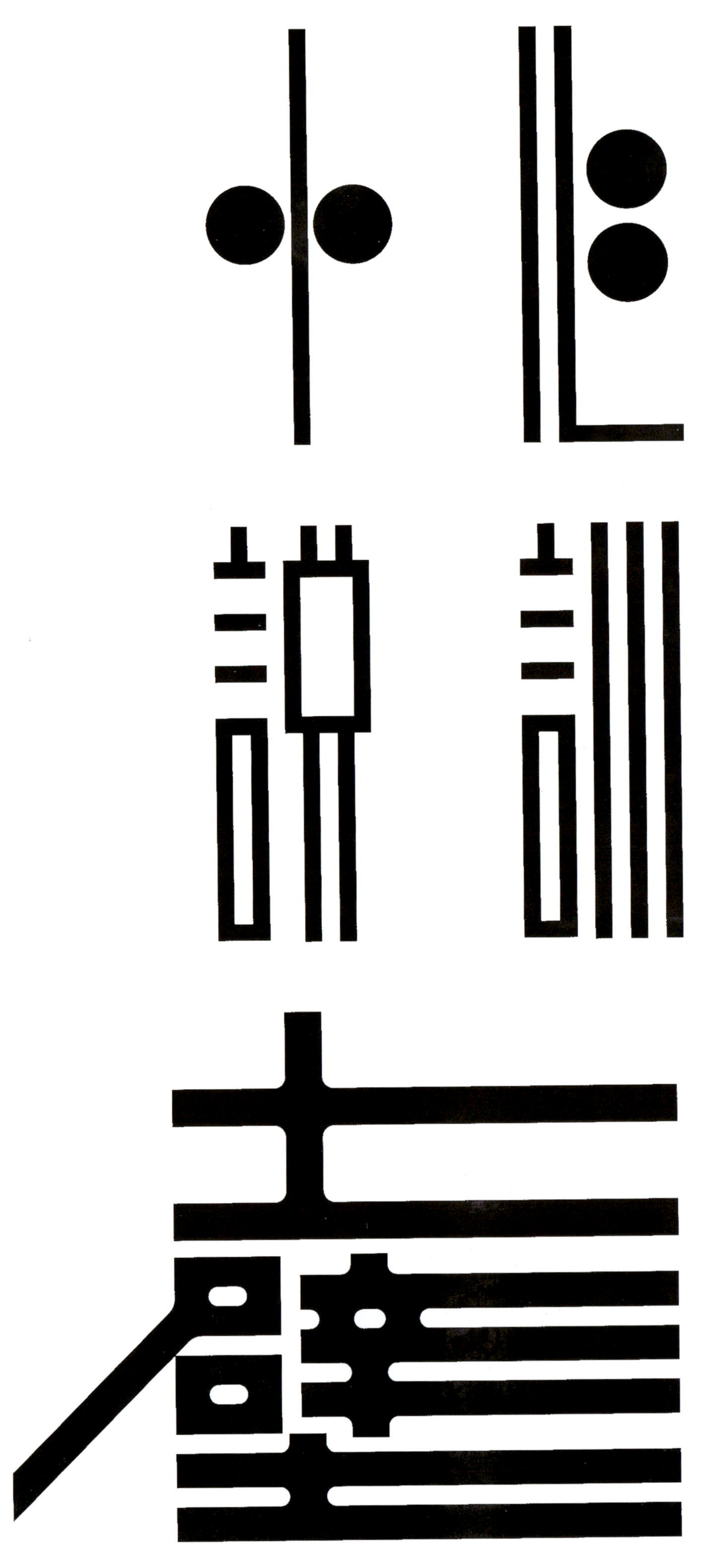

A moving card for Royse Sanders & Amfitheatreof, a company specializing in arts-related projects. Set in a bold condensed sans-serif font, the letters 'RS&A' morph into 'have', which then morphs again into 'moved'. As well as changing shape, the letters go through a spectrum of colours along their journey.

Created for an invite to an exhibition organized by *Creative Review* magazine, each character morphs into the next. The text reads: 'We are your future 15 stars of tomorrow in advertising and design.' The characters move around the page as well as morphing and changing colour, which gives the invite a dynamic energy.

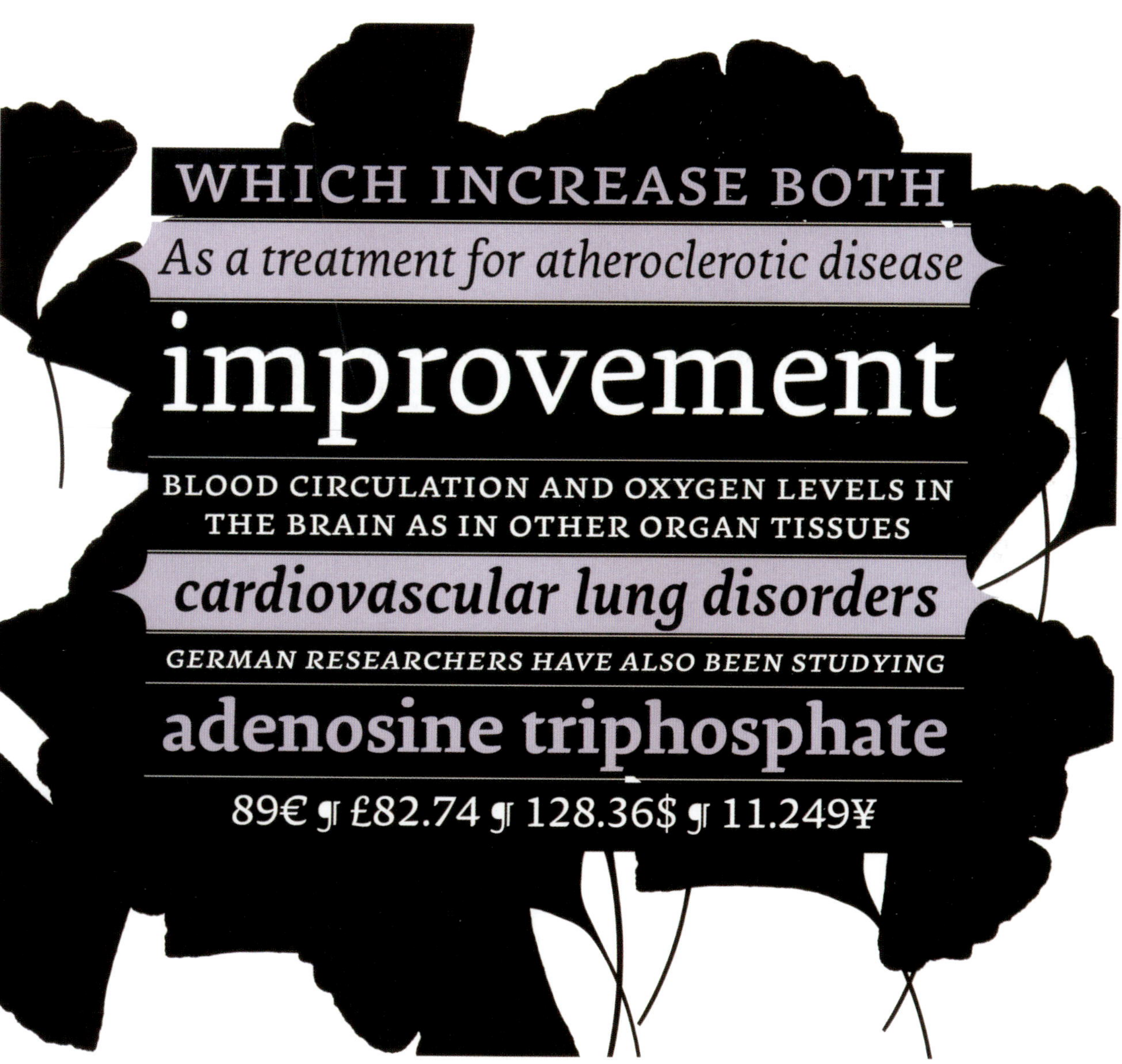

Beginning life as an identity for the typefoundry Dalton Maag, the DaMa logo has been extended into a full typeface. The fat rounded letterforms have a retro quality which is brought up to date by the reduced, minimal shape of each character. The font is only available in one heavy weight, which is generated in both a solid and an outline variant.

Produced for a typefoundry, this brochure highlights a series of bespoke corporate fonts created by it. Large fragments of the fonts are shown against photographic backdrops relevant to each case study. The typography is interwoven with screen grabs of the fonts under construction.

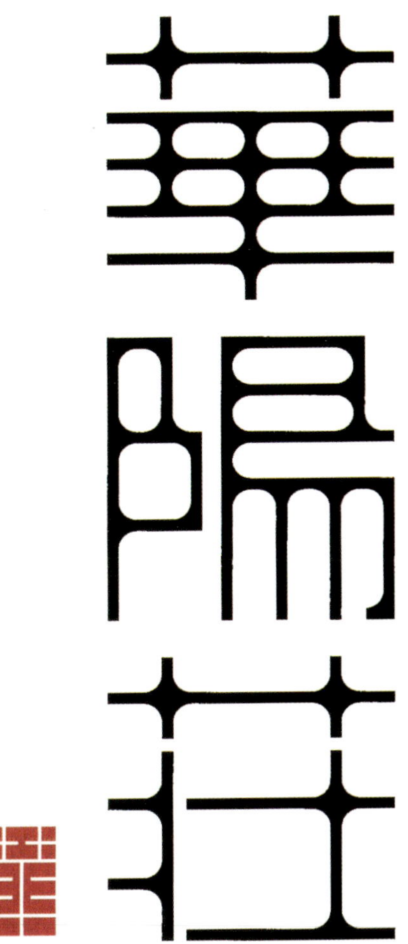

華陽莊

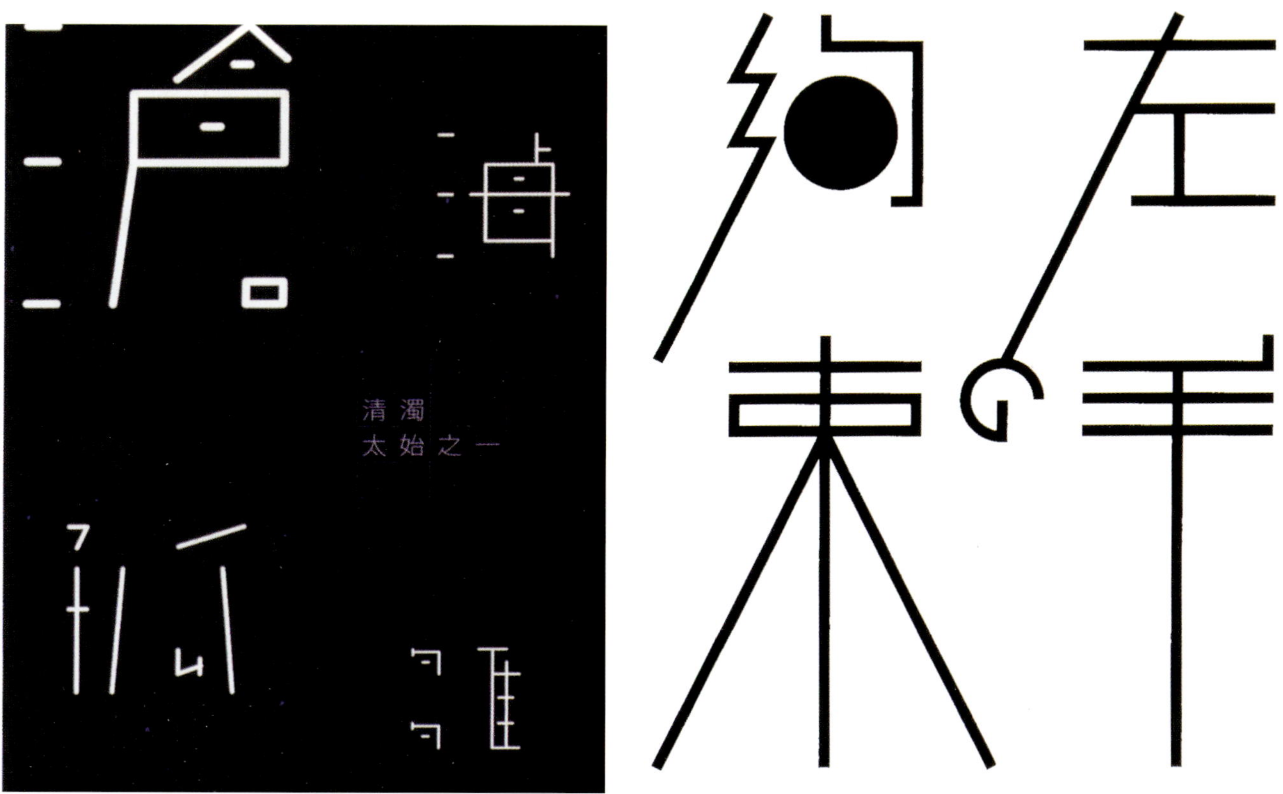

戶
清濁
太始之一
約
左
束
羊

This annual report for the mobile phone company Nokia combines two different formats that carry different levels of information. The large French-folded pages feature a series of informal photographs of people printed onto a thin stock; a series of large typographic statements, some of which are printed in mirror image on the inside of the French-folded pages, are just legible through the light paper. A sequence of smaller pages are bound into the bottom of the brochure, printed on an uncoated stock, and feature the more conventional annual report information.

As part of the extensive global branding exercise undertaken by Nokia, this range of carrier bags was developed using a purely typographic solution. A different bag has been created for every country in which the company operates. The country name and international dialling code are printed in the palette of Nokia brand colours around the bag, allowing only fragments of the type to be read on any one individual bag face.

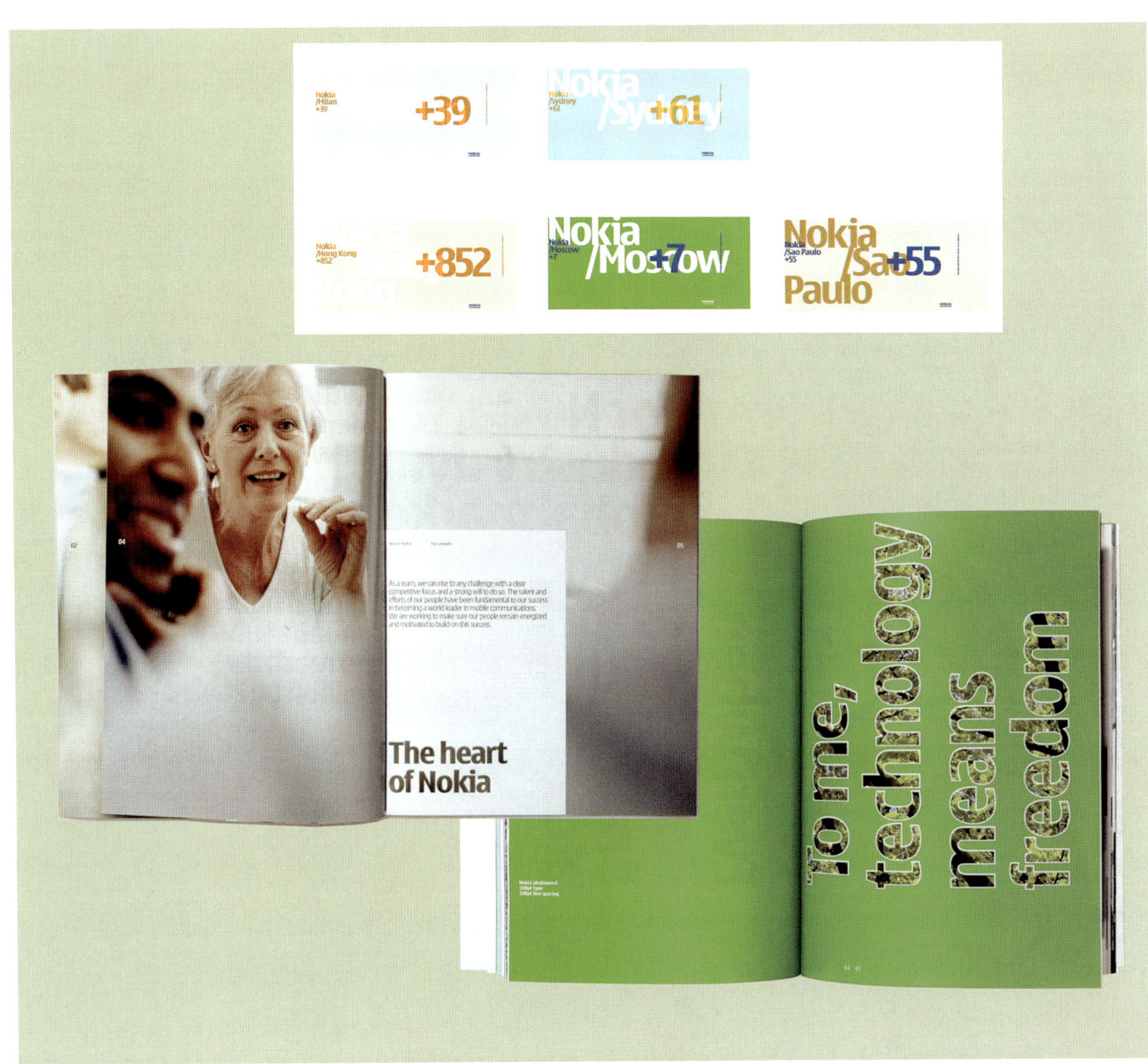

Since the people working with Taxi Magazine is traveling and moving people, situated in different places around the world, the stationary needed to be usable in every situation and reflecting the diversity of Taxi. Therefor, it is designed as one simple sticker. Containing logotype and contact details, ready to apply on all kinds of papers, depending on taste and situation.

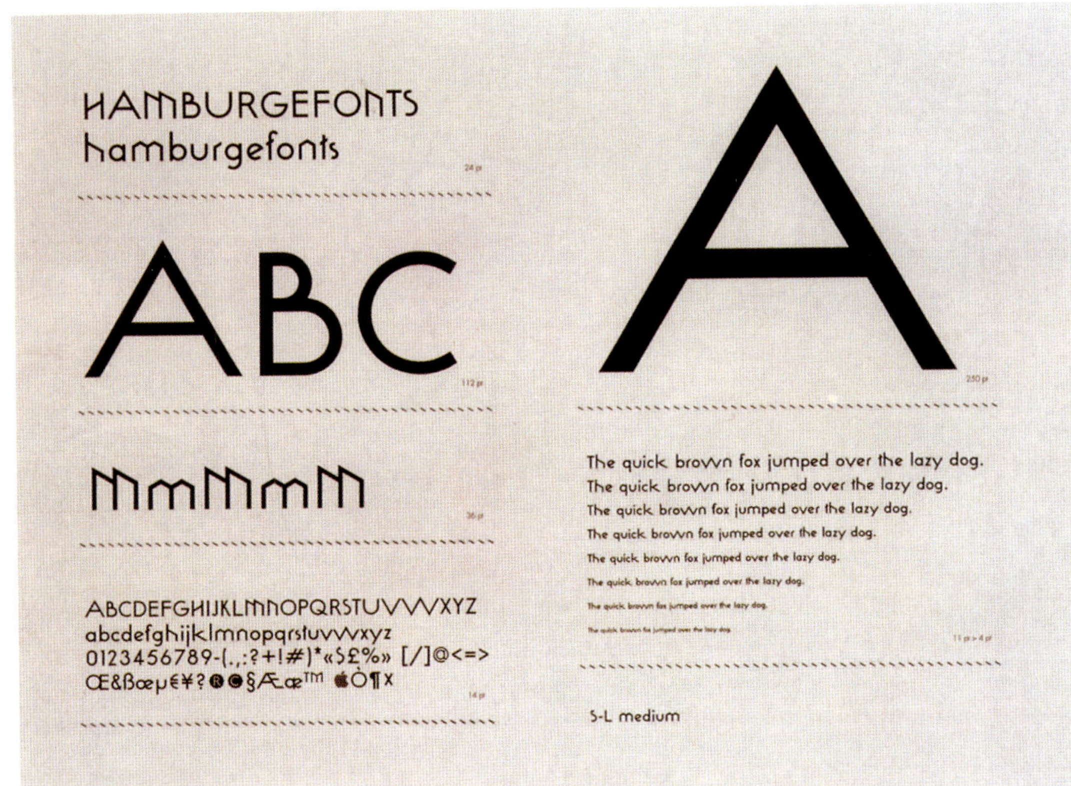

HAMBURGEFONTS
hamburgefonts
24 pt
ABC
112 pt
MmNmM
36 pt
ABCDEFGHIJKLMNOPQRSTUVVVXYZ
abcdefghijklmnopqrstuvvvxyz
0123456789-(.,:?+!#)*«$£%» [/]@<=>
Œ&ßœµ€¥?®©§Æœ™ Ó¶x
14 pt
A
250 pt
The quick brovvn fox jumped over the lazy dog.
The quick brovvn fox jumped over the lazy dog.
The quick brovvn fox jumped over the lazy dog.
The quick brovvn fox jumped over the lazy dog.
The quick brovvn fox jumped over the lazy dog.
The quick brovvn fox jumped over the lazy dog.
The quick brovvn fox jumped over the lazy dog.
The quick brovvn fox jumped over the lazy dog.
11 pt > 4 pt
S-L medium

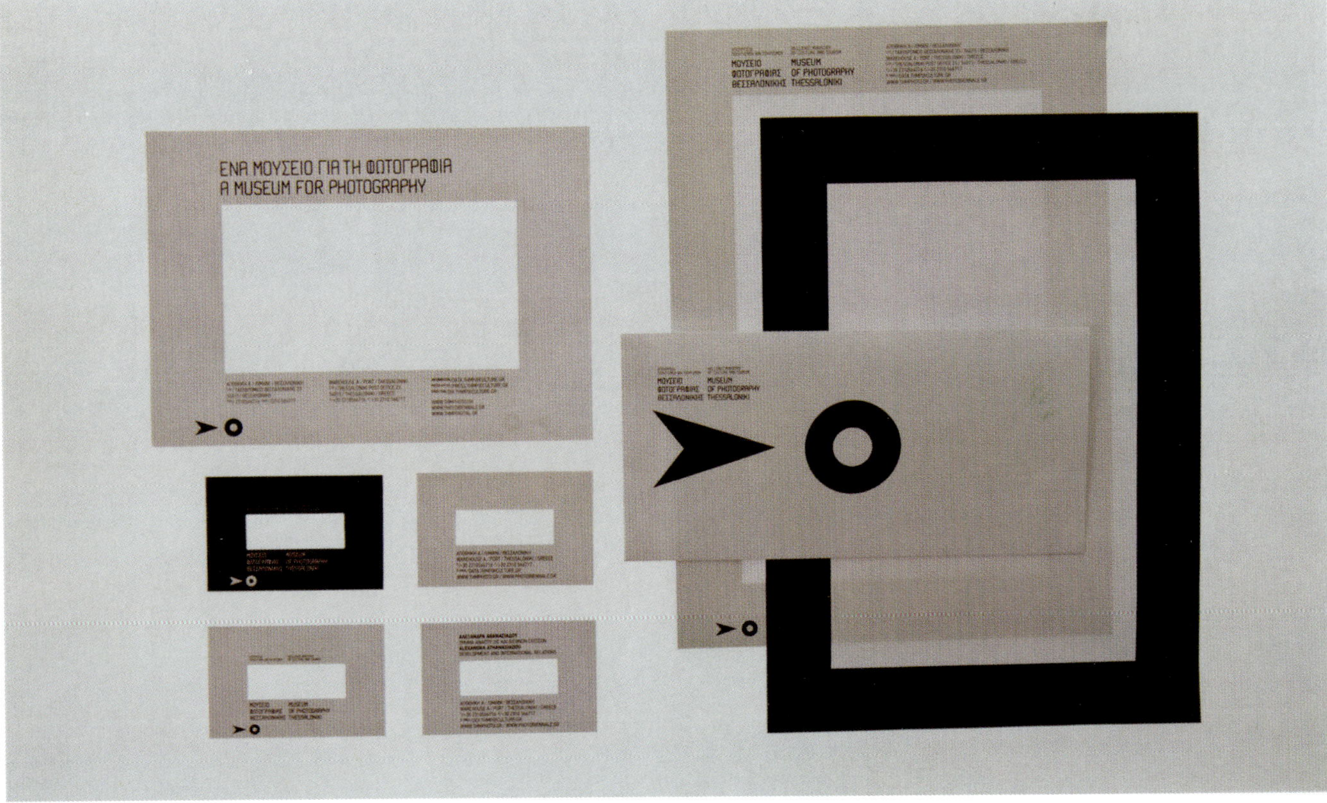

ΕΝΑ ΜΟΥΣΕΙΟ ΓΙΑ ΤΗ ΦΩΤΟΓΡΑΦΙΑ
A MUSEUM FOR PHOTOGRAPHY
ΜΟΥΣΕΙΟ   MUSEUM
ΦΩΤΟΓΡΑΦΙΑΣ   OF PHOTOGRAPHY
ΘΕΣΣΑΛΟΝΙΚΗΣ THESSALONIKI

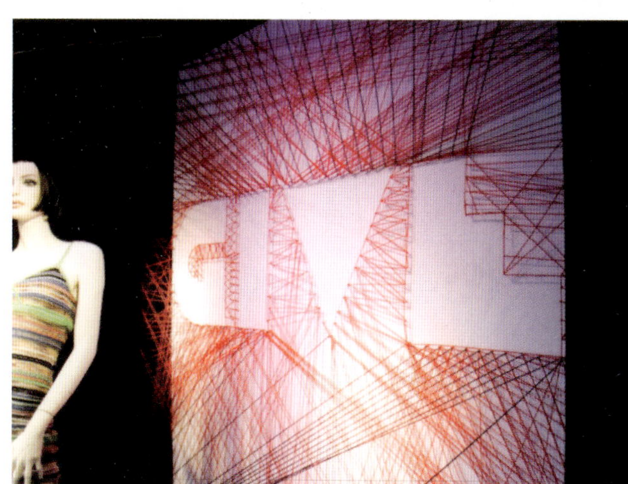

Believe

Using the empty bowels of a cargo ferry as a background, this TV ident for MTV Networks International features digitally generated typography that constructs itself from abstract modules. Fragments fly into view and morph into three-dimensional letterforms.

A simple folded sheet of high-gloss black paper is enlivened by the application of silver glitter to the custom-drawn typography. The geometric typography on this New Year's card from the London-based design consultancy is constructed from just a circle and a straight line. The design combines a retro Seventies disco aesthetic with a clean modern sensibility.

Based on one of the designer's own typeface designs, these New Year's resolutions have been sprayed onto the studio wall to serve as a semi-permanent reminder for the rest of the year.

typeface

BD Doomed
SquareUp

typeface family
BD Doomed SquareUp

designer
MBrunner

foundry/supplier
büro destruct

country of origin
Switzerland

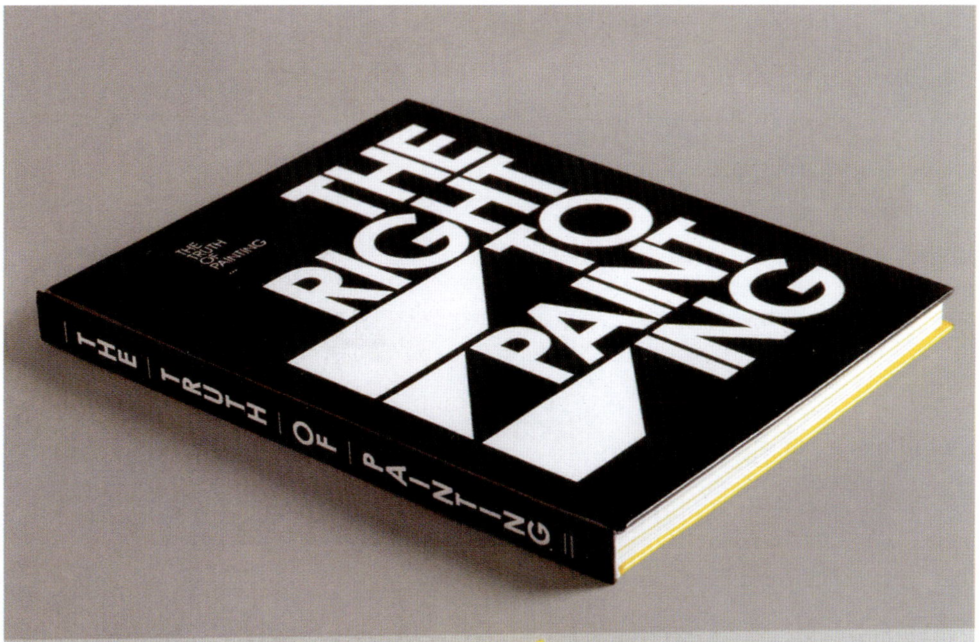
THE RIGHT TO PAINT ING
THE TRUTH OF PAINTING

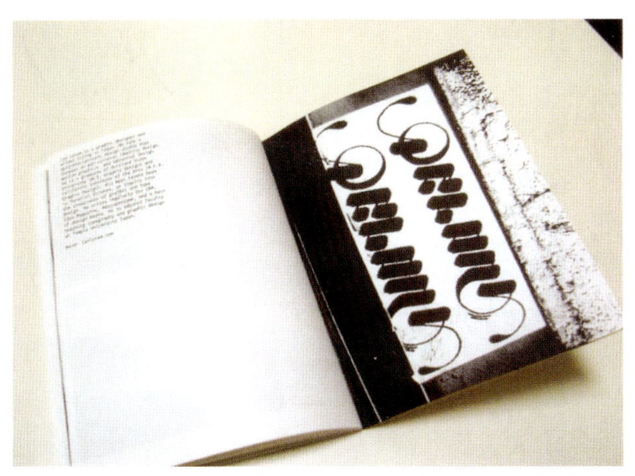

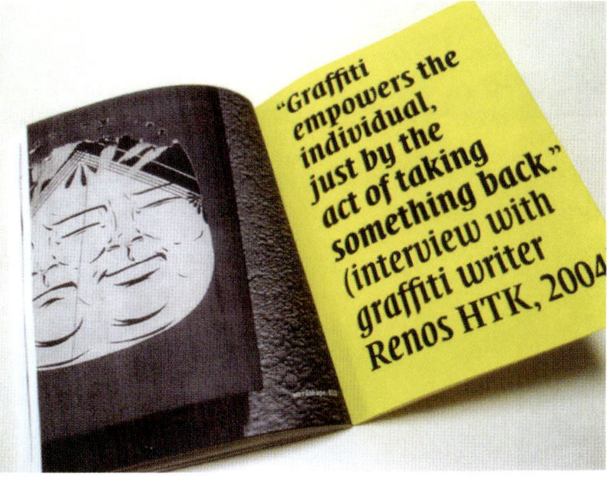

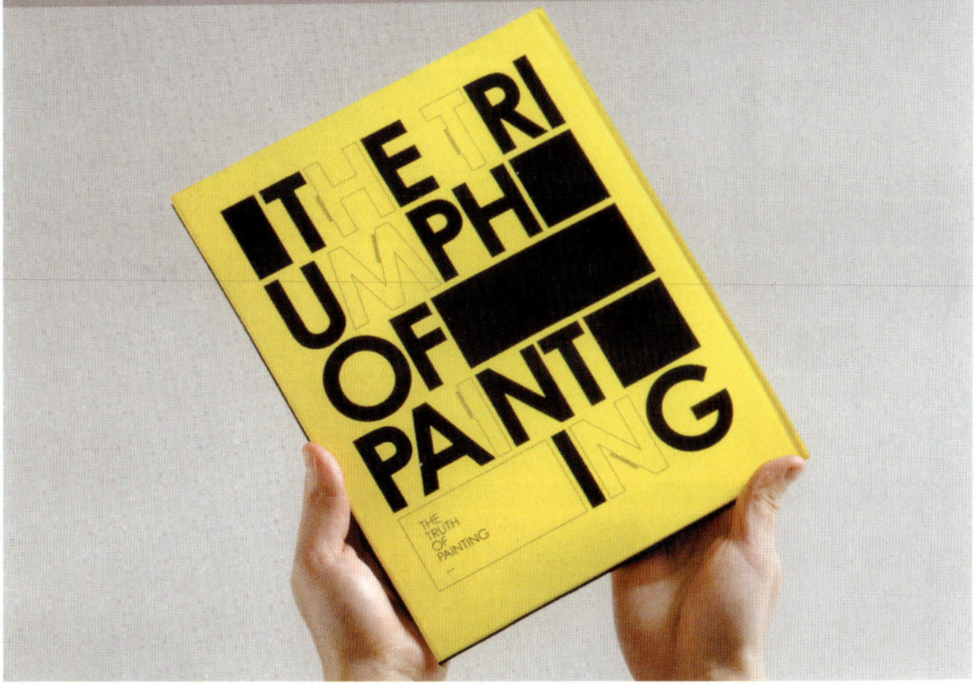
THE TRUTH OF PAINTING

"Graffiti empowers the individual, just by the act of taking something back."
(interview with graffiti writer Renos HTK, 2004

jan
de
cock
WERK No.13
Spring/Summer 2008
Recorded and photographed by
Kirby Koh at the Tate Modern in London

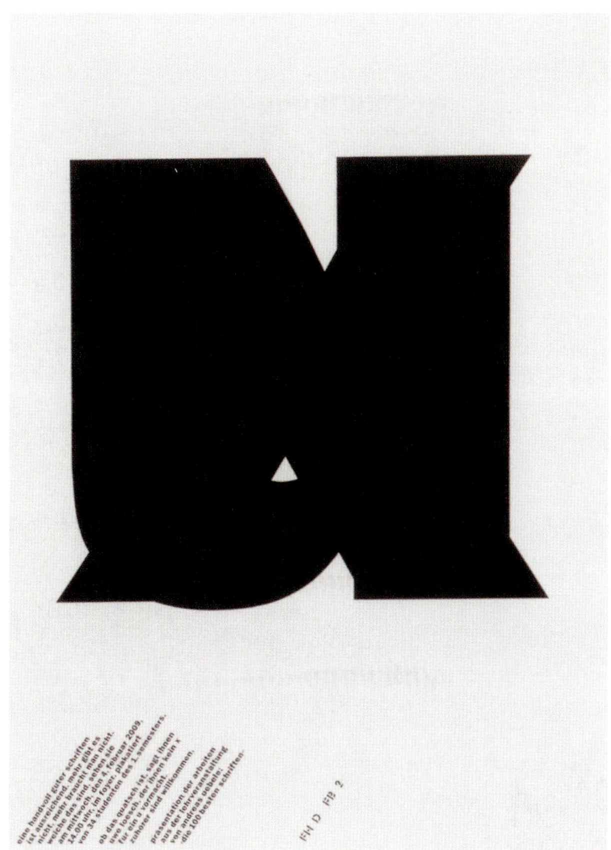

271 Poster: TD. D. Mirko Ilic/ D. Jee-eun Lee/
CL. AIC Foundation/ PT. Custom-made for the project

272_Poster: TD. Andreas Uebele/
CL. Dusseldorf University of Applied Sciences. Department of Design/ PT. Franklin Gothic

IMMENSTADT ALLGÄU
EBERL
1859
BERGKÄSE
VON BERND
E
03

IMMENSTADT
EBERL
SEIT 1859
BERGKÄSE
BUTTERKÄSE
EMMENTALER
TILSITER
ALLGÄUER KÄSESORTEN
03

überreicht von
STEFANIES TILSITER
EBERL
SINCE 1859
50 %
FETT I. TR.
leicht säuerlich, würzig

MELKEN